TOLERANCE STACK-UP ANALYSIS

[for Plus and Minus and Geometric Tolerancing]

Second Edition

©James D. Meadows

Copyright © 2010 James D. Meadows
Second Edition

Earlier Edition: Copyright 2001 James D. Meadows

ALL RIGHTS RESERVED including those of translation. No part of this publication may be reproduced, stored in a retrieval system, or transmitted in any form or by any means—graphic, electronic, mechanical, including photocopying, recording, taping, Web distribution, or otherwise—without written permission of the author.

Published and distributed by:
James D. Meadows & Associates, Inc.
170 E. Main, D-137
Hendersonville, TN 37075
 Phone: (615) 824-8644
 FAX: (615) 824-5262
 www.geotolmeadows.com
 JDMeadows@geotolmeadows.com

ISBN: 978-0-9714401-4-x
Printed in the United States of America

3 4 5 6 7 8 9 10 Printing

No liability is assumed by the publisher James D. Meadows & Associates, Inc., nor its author, with respect to the use of the information contained herein. Information contained in this work has been obtained from sources believed to be reliable. While every precaution has been taken in the preparation of this book, neither James D. Meadows & Associates, Inc., nor its author guarantee the accuracy or completeness of any information published herein and neither James D. Meadows & Associates, Inc., nor it author shall be responsible for any errors, omissions, or damages arising out of use of this information. This work is published with the understanding that James D. Meadows & Associates, Inc., and its author are supplying information but are not attempting to render engineering or other professional services. The publisher and its author shall not be liable for any special, consequential, or exemplary damages resulting in whole or part, from the readers' use of, or reliance upon, this material.

FOREWORD

Tolerance Stack-Up Analysis is a subject that has been around for many decades, practiced on plus and minus toleranced parts, done by engineering professionals and programmed into software. At present, many tolerance analysts are asked to perform stack-ups to determine interference possibilities, maximum gaps, and minimum housing requirements for assemblies as well as to calculate a wide variety of other unknowns. In order to perform a tolerance stack-up analysis on today's complex assemblies, a thorough understanding of Geometric Dimensioning and Tolerancing is needed. But even a masterful understanding of GD&T is not enough to successfully calculate many of the assembly problems, material weaknesses and other elusive unknowns one needs to find. It requires a knowledge of the product, a sound methodology, a correct visualization of the parts in the assembly pushed to create the worst situation and, most of all, good sound judgment of which dimensions and tolerances are and are not factors.

Tolerance Stack-Up Analysis is an entirely different way of looking at assemblies. Unlike Geometric Dimensioning and Tolerancing practices, which analyze worst case boundaries, but best case assembly conditions, Tolerance Stack-Up Analysis calculates difficulties caused by both worst case boundaries and worst case assembly conditions. GD&T usually assumes the parts will be assembled optimally, so that even if part features are produced at their worst size, form, orientation and location, they have been dimensioned and tolerance in such a way that they will still fit together. Tolerance Stack-Up Analysis looks at it differently. It not only considers the parts will be produced at their worst case size, form, orientation and location, but that they will also be pushed and sometimes even rocked into their most unfriendly assembly conditions. It determines what the worst case assembly conditions would be if we tried to push, rock and rotate the parts in the assembly to their extremes. In many cases, this is not the most logical approach, but it certainly is the worst of all assembly condition.

This book assumes a moderate level of GD&T knowledge has already been achieved by the reader, but it begins with a thorough explanation of all of the simple (and some not so simple) preliminary procedures one must perform to begin a successful Tolerance Stack-Up Analysis. It teaches one to convert a dimension from a variety of tolerancing presentations to equal bilateral tolerancing. It then proceeds to show how to calculate inner and outer boundaries using geometric tolerances and convert them to equal bilaterally toleranced dimensions. It shows how to do Tolerance Stack-Up Analysis on some very simple plus and minus toleranced parts and then proceeds to extend these procedures to geometrically toleranced parts. It covers fixed and floating fastener assemblies, showing a logical, easy to follow approach with some very simple rules. It explores areas that include the most important components of the analysis that include following the procedures to the letter and including in that procedure the judgment of the analyst. This is critical to the analysis and why software programs have so much trouble analyzing assemblies and actually arriving at correct answers. Following a procedure **without questioning** every inclusion or exclusion of a tolerance, to reason out whether or not it is a factor in finding the answers one is looking for and also asking the question, "What have I forgotten to include?", is a sure way to get it all wrong. Tolerance analysis is a house of cards, where if you stack one card wrong, the entire house falls; one missed factor, one wrong factor, one wrong judgment, even an addition or subtraction error results in the entire analysis failing.

There are some aspects of Tolerance Stack-Up Analysis that other books do not cover, but this one does explore. Statistical Analysis and Statistical Tolerancing are covered thoroughly in Chapter 11. Trigonometric aspects and contributors are discussed in Chapter 10 and can be so time consuming as to be unrealistic to take on without computer programs to assist the analyst. But trigonometric factors and variables that are found through the use of proportions and algebraic equations are calculable and should not be ignored by the analyst. Software can be a valuable assistant in doing stacks, but should never take the place of the good, sound judgment of the analyst. It is critical that the user of tolerance analysis software performs tests of the program to determine what mistakes it is prone to make. The question is not, "Will it make mistakes?". Without question mistakes will be made. The questions are, "What mistakes does it make?", and "Are those errors acceptable to you?" or "Can you compensate for the errors it makes?". The procedures explained in this book will give the reader the basis to perform any simple Stack-Up Analysis and to assess procedures being performed by programs meant to do more complicated analyses.

I would like to thank Don Brown, who worked for Chrysler Corporation, and Anthony Teresi, who was at Novellus Corporation. Don badgered me to do a course on Tolerance Stack-Up Analysis and it seems that, these days, I never do courses without first writing a book on the topic. Anthony gave me the basis for Chapter 7, which was the first chapter I wrote. Our conversations about the topic were centered on the sobering premise that software programs can't yet think. We determined that you can perform a great process absolutely correctly and get an answer that is either entirely wrong, unlikely or defies the laws of what is physically possible. That is precisely the type of problem that interests me. It lets me know that the minds of men will not be easily replaced by computers. Until we learn to program in product knowledge, reason and logic, there will be a need for engineers, technical personnel and sentient beings in general. Ultimately, it is up to us to assess the results of these computer analyses. I like the way that feels. Computers are here to assist us, not to replace us.

The procedures in the first six chapters of this book came from the work of Carl Lance. He was my teacher and one of the most brilliant men I have ever met. Without his guidance, example and years of patient explanations, this book and the others I have written would never have come into existence. I consider this book as merely an expansion of his work.

As always, I remain indebted to all those who helped construct this book, specifically to Michael Gay of Nashville CAD, Inc., for his great illustrations and Jeannie Winchell for overseeing and coordinating the entire project.

This revision makes Chapter 5 easier to understand, includes more discussion of what can go wrong in Chapter 7, shows the right answers and how to tell right from wrong and includes more on statistical analysis in Chapter 11. Other chapters have been changed to make them read smoother or to add or subtract information that I felt would ease the process of learning. The terminology in this revision has been updated to comply with ASME Y14.5-2009.

James D. Meadows
December 2009

Tolerance Stack-Up Analysis
TABLE OF CONTENTS

Chapter #	Page
#1 THE BASICS	
Topics	1
•Where to begin a stack	2
•Designating positive and negative routes	2
•What are you calculating?	3
•What dimensions are factors	3
•How to push the parts to create the worst case	4
•Which geometric tolerances are and are not factors	4
•Finding the mean	5
•Calculating boundaries for GD&T, MMC, LMC and RFS Material Condition modifiers	5
•Mean boundaries with equal bilateral tolerances	6
Exercises	13
#2 STACK-UP ANALYSIS OF AN ELEVEN PART ASSEMBLY USING PLUS AND MINUS TOLERANCING	
Topics	16
•The calculations	17
•The loop analysis chart	21
•The numbers analysis chart	21
•Finding MIN and MAX gaps	23
Exercises	24
#3 VERTICAL vs. HORIZONTAL LOOP ANALYSES FOR FEATURES OF SIZE	
Topics	27
•Where to start and end	29
•Graphing the loop	29
•Minimum and maximum gap analysis	32
Exercises	33
#4 ASSEMBLIES WITH PLUS AND MINUS TOLERANCES	
Topics	36
•Multiple dimension loops	37
•Positive and negative values	40
•Airspace vs. interferences	41
Exercises	43
#5 FLOATING FASTENER FIVE PART ASSEMBLY ANALYSIS	
Topics	46
•Resultant conditions	49
•Virtual conditions	
•Inner and outer boundaries	49
•Mean boundaries	49
•Converting to radii	49
•Mixing widths and diameters	53

•Complex loop analyses with geometric dimensioning and tolerancing	53
Exercises	**60**

#6 FIXED FASTENER ASSEMBLIES
Topics	**65**
•Calculating overall minimum and maximum assembly dimensions	66
•Mixing slots, tabs, holes and shafts	66
•Calculating minimum and maximum gaps within the assembly	69
•Projected tolerance zones and total runout as factors	86
•Determining if geometric tolerances are a factor	86
•Ruling out features and patterns as factors	86
Exercises	**89**

#7 A RAIL ASSEMBLY
Topics	**99**
•Threaded features	101
•Multiple geometric controls	101
•Projected tolerance zones	101
•How to tell a right answer from a wrong answer	**102**
•Theoretically vs. physically worst case possibilities	103
•When logic becomes an integral step	103
•Calculating interference	109
•Factoring in assembly conditions	117
•Maximum wall thickness vs. minimum airspace for assemblies	119
•Gaps with and without perpendicularity as a factor	127
Exercises	**131**

#8 SINGLE-PART ANALYSIS
Topics	**145**
•Two-single segment positional controls	146
•Creating envelopes of perfect orientation at MMC	146
•Switching datum reference frames and accumulating geometric tolerances	147
•Datum features at MMC (pattern shift)	147
•Flatness	156
•Envelopes of perfect form at MMC	156
•Datum planes vs. datum features	156
•Profile tolerances	157
•MIN and MAX axial separation	158
•Separate requirements and accumulating tolerance	165
Exercises	**169**
•Tolerances in degrees; Trigonometric function introduction	174
•Composite positional tolerancing	177

#9 FIVE PART ROTATING ASSEMBLY ANALYSIS
Topics	**178**
•Position	180
•Perpendicularity	180
•Parallelism	180
•Profile	180
•Flatness	180

•Threaded holes with projected tolerance zones	181
•Mounted screws	181
•Part to part analysis (from two parts to an infinite number of parts	181
•Positional coaxiality	181
•Runout	182
•Total runout	182
•Simplifying a complex assembly	186
•Determining assembly housing requirements	190
•Radial clearance MIN calculations	190
Exercises	192

#10 TRIGONOMETRY AND PROPORTIONS IN TOLERANCE STACK-UP ANALYSIS

Topics	199
•Rocking datum features	200
•Constructing a valid datum	200
•Consideration of differing orientations from measurement to assembly	200
•An in-depth assembly analysis using proportions and trigonometric functions	201
•Vertical stacking as it effects horizontal housing requirements	202
•When stacked parts are not flat or parallel	202
•Using proportions and trigonometry to calculate fit conditions beyond the GD&T formulae	202
•Formulae to calculate worst case fit conditions when trigonometry is a factor	205
•Computer programs vs. a personal analysis	208
Exercises	212

#11 THE THEORY OF STATISTICAL PROBABILITY

Topics	214
•Gaussian Frequency Curve	216
•Standard Deviations	216
•Plus or Minus 3 Sigma	216
•Root Sum Square Formula	216
•Steps to Calculate and Apply Statistical Tolerances	216
•Statistical Tolerancing Applied to Plus and Minus Toleranced Assemblies	219
•Statistical Tolerancing Applied to Geometric Toleranced Assemblies	221
•When Best to Allow Statistical Tolerances and When it Should Not Be Allowed	224
•The Logic of Statistical Tolerancing	224
•Modifying the Root Sum Square Formula with a Safety/Correction Factor	224
•Reintegrating the Statistical Tolerance into the Assembly	225
•A Simpler Way?	228
•More Statistical Formulas and Symbols	229
•Glossary of Statistical Terms	231
Exercises	234

ANSWERS	238
Chapter 1	239-242
Chapter 2	243-244
Chapter 3	245
Chapter 4	246
Chapter 5	247-249
Chapter 6	250-257
Chapter 7	258-264

Chapter 8   265-285
Chapter 9   286-287
Chapter 10   288
Chapter 11   289-296

APPENDIX A   A-1
Course Materials Written by James D. Meadows   **A-2**
On-Site Courses Available   **A-3**

Chapter 1

THE BASICS
- Calculating Mean Dimensions with Equal Bilateral Tolerances
- Calculating Inner and Outer Boundaries
- Virtual and Resultant Conditions

• Lesson Objectives:

In Chapter 1, you will learn to:
- Identify the factors pertinent to the stack-up analysis.
- Designate positive and negative routes.
- Position each part in the assembly for a worst-case analysis.
- Find a mean and its equal bilateral tolerance for limit dimensions, unilateral and bilateral toleranced dimensions.
- Calculate inner and outer boundaries for geometric tolerancing, then convert to a mean with an equal bilateral tolerance.

TOLERANCE STACK-UP ANALYSIS

Main Rules

1) Start at the bottom (of the part or gap) and work to the top.

... or

Start at the left (of the part or gap) and work to the right.

2) Stay on one part until it is exhausted, then jump to another-- not back and forth.

3) Left is negative (-)

... &

Down is negative (-)

Right is positive (+)

... &

Up is positive (+)

4) Always take the shortest route (so you won't accumulate unnecessary and incorrect tolerance).

Chapter 1
The Basics

Tolerance Stack-Up Analysis (Gap Analysis, Loop Diagrams or Circuit Analysis) is (among other things) the act of calculating minimum and maximum airspace or material interferences in assemblies. The process can be broken down into steps.

Step 1
- The first step is to **identify what requirement is under test**. For example, if at a certain place in the assembly you want to determine that no interference is possible, you set your procedure to calculate **the gap that must be equal to or greater than zero (no material interference)**.

Step 2
- The second step is to **identify all dimensions that contribute to the gap** you are calculating and convert them to equal bilaterally toleranced dimensions (if they are not already toleranced as equal bilateral).

Step 3
- The third step is to **assign each dimension a positive or negative value**. Radial stacks (going up and down) start at the bottom of the gap and end up at the top of the gap.
  - **Down is negative.**
  - **Up is positive.**

Stacks that go left and right in the assembly start at the left side of the gap and end up at the right side of the gap.
  - **Left is negative.**
  - **Right is positive.**

Remember, you are working one part at a time, so do one part's significant features before jumping to the next part's significant features. In this way, an infinite number of parts will be little more difficult than a couple of parts.

Step 4: ... another basic rule
Remember, one set of mating features from part to part creates the variable you are searching for. Whether it is the minimum gap, maximum gap or maximum overall assembly dimension, one set of mating features creates it. So, although multiple routes may have to be examined to find this contributing set of features, only one set creates the worst case, from one part to the next.

It is a mistake to follow one route from one set of mating features (holes and shafts), then continue the same route through another set on the same part. One of these sets creates the smallest or biggest gap or maximum overall dimension. Once you find out which it is, the others become non-factors in the analysis.

Using more than one set of mating features from one part to the next (from the same two parts) will most likely give the analyst a wrong answer and never gives a correct answer that couldn't have been found using the one correct set of mating holes and shafts. Still, tolerances from other features may contribute to the critical set you are using. Examples of this are: a) when datum features are referenced at MMC, and b) when more than one set of datum features come into play.

Step 5: . . . and, another basic rule

At times, a single feature or a pattern of features has two geometric tolerances--for example, both a position and a perpendicularity tolerance. In these situations, the analyst must determine which, if either, is the contributing factor to the unknown being calculated. It is also possible that neither geometric tolerance is a factor but, instead, only the size dimensions and their tolerances are factors. The analyst must deduce which factors are pertinent through sketches and deductive reasoning. The judgment of the analyst is critical in these determinations.

To begin, **add all positive and negative dimensions** which will calculate your mean gap. If the mean gap is a negative number, your requirement of no material interference has already been violated.

Then, often we must **add the sum of the equal bilateral tolerances (1/2 the total tolerance) to the mean dimension** to determine the maximum gap. Then subtract the sum of the equal bilateral tolerances (1/2 the total tolerance) from the mean dimension to calculate the minimum gap. Again, any negative final values for minimum or maximum gaps indicate material interference. Maximum gaps are equal to maximum clearance (or in the case of interference fits, minimum interference). Minimum gaps are equal to the minimum clearance (or in the case of interference, the maximum interference). It is also important to mentally shove all features and parts in the direction that will create the minimum or maximum gap or interference you are calculating. This is to allow your routes to always pass through material. You do not want to jump over an airspace (without numbers to lead you across) unnecessarily in your calculation. You want to position the features of the pieces against each other so that you will get the extremes and make clear to you the correct path and either a positive or a negative designation for each number.

Step 6: If your assumptions are wrong, your answer is wrong.

IMPORTANT RULES:

Correct routes give the largest MIN GAP and the smallest MAX GAP.

Run multiple routes to determine (for example) MIN GAP. The route that gives you the largest MIN GAP is the one that is correct, because it is the one that aligns (and stops the gap from being smaller) or hits first (and stops the gap from being smaller).

MAX GAP: The route that gives you the smallest MAX GAP is the one that is correct, because it is the one that hits first.

These routes assume your "one line" is the correct place where one part contacts another.

Always take the shortest route. It is easy to get a wrong answer if your route is not the shortest because it will accumulate error (tolerances) that should not be included in your calculation.

Finding the mean:

First we must understand a few concepts of tolerance analysis. One is that there is no difference between an equal, unequal or unilaterally toleranced dimension. There is no difference between a limit dimension and a plus or minus toleranced dimension. They all have extremes and they all have means. So, the first thing we will do is change any dimension to an equal <u>bilaterally toleranced</u> dimension.

For example:

$$\emptyset 22 = \begin{array}{r} 22 \\ +20 \\ \hline \emptyset 42 \end{array} \quad \& \quad \begin{array}{r} 22 \\ -20 \\ \hline \emptyset 2 \end{array}$$

So;

$$\frac{42}{2} = \emptyset 21 \quad \& \quad \frac{\emptyset 2}{2} = \emptyset 1$$

Therefore, 20 - 22 as a limit dimension is $\emptyset 21 \pm 1$ as an equal bilateral toleranced dimension and $\emptyset 50 \begin{array}{c} +1 \\ -3 \end{array}$ is

$$\begin{array}{r} 50 \\ +1 \\ \hline \emptyset 51 \end{array} \quad \& \quad \begin{array}{r} 50 \\ -3 \\ \hline \emptyset 47 \end{array}$$

$$\emptyset 51 = \begin{array}{r} 51 \\ +47 \\ \hline 98 \end{array} \quad \& \quad \begin{array}{r} 51 \\ -47 \\ \hline 4 \end{array}$$

So;

$$\frac{98}{2} = \emptyset 49 \quad \& \quad \frac{4}{2} = 2$$

Therefore, $\emptyset 50 \begin{array}{c} +1 \\ -3 \end{array}$ as an unequal bilateral toleranced dimension is:

$\emptyset 49 \pm 2$ as an equal bilateral toleranced dimension.

Boundaries:

The concepts of boundaries generated by the collective effects of size and geometric tolerances are often referred to as simply <u>inner</u> boundaries and <u>outer</u> boundaries, but they are also called virtual conditions and <u>resultant</u> conditions. Any of these boundaries that are constant are defined as virtual conditions and the "worst case" of non-constant boundaries are resultant conditions.

Feature control frames that use maximum material condition symbols after the geometric tolerances generate constant (virtual condition) boundaries on the inner side for holes and on the outer side for shafts. Any feature control frames that use least material condition symbols after geometric tolerances generate constant (virtual condition) boundaries on the outer side for holes and on the inner side for shafts.

In all of these conditions, the <u>resultant</u>, or "<u>worst case</u>" for the non-constant boundaries are generated on the opposite side of the virtual conditions. The regardless of feature size concepts, when used on holes or shafts in the feature control frames, generate only "worst case boundaries" (resultant conditions) for the non-constant boundaries.

FIGURE 1-1 [Holes-MMC Concept]

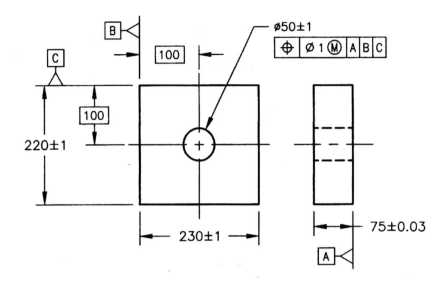

Constant vs. Non-Constant Boundaries [FIGURE 1-1]

HOLE

Size	Geo. Tol.		Virtual Condition (constant boundary)
Ø49	- 1	=	Ø48) → worst case inner boundary
Ø50	- 2	=	Ø48)
Ø51	- 3	=	Ø48)

HOLE

Size	Geo. Tol.		Resultant Condition (worst case boundary)
Ø49	+ 1	=	Ø50
Ø50	+ 2	=	Ø52
Ø51	+ 3	=	Ø54 → worst case outer boundary

Creating an Equal Bilateral Toleranced Dimension from Virtual and Resultant Conditions (MMC Concept Hole • FIGURE 1-1)

```
  Resultant Condition Hole   =  Ø54
+ Virtual Condition Hole     =  Ø48
  Sum                        =  Ø102

  Resultant Condition Hole   =  Ø54
- Virtual Condition Hole     =  Ø48
  Difference                 =  Ø6
```

Then:

$$\frac{102}{2} = 51 \quad \& \quad \frac{6}{2} = 3$$

So, Ø51±3 is an equal bilateral expression of the dimension and its tolerance.

FIGURE 1-2 [LMC Concepts]

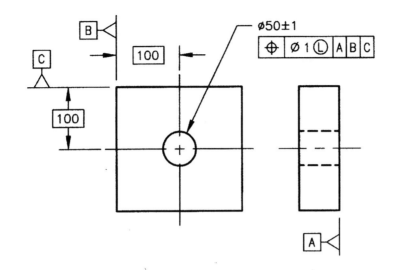

SKIP

Constant vs. Non-Constant Boundaries [FIGURE 1-2]

HOLE

Size	Geo. Tol.			Virtual Condition (constant boundary)
Ø51	+	1	=	Ø52 → worst case outer boundary
Ø50	+	2	=	Ø52
Ø49	+	3	=	Ø52

HOLE

Size	Geo. Tol.			Resultant Condition (worst case boundary)
Ø51	-	1	=	Ø50
Ø50	-	2	=	Ø48
Ø49	-	3	=	Ø46 → worst case inner boundary

Creating an Equal Bilateral Toleranced Dimension from Virtual and Resultant Conditions [FIGURE 1-2]

```
  Resultant Condition Hole    =    46
+ Virtual Condition Hole      =    52
         Sum                  =    Ø98

  Virtual Condition Hole      =    52
- Resultant Condition Hole    =    46
         Difference           =     6
```

Then:

$\dfrac{98}{2} = 49$ & $\dfrac{6}{2} = 3$

So, Ø49±3 is an equal bilateral expression of the dimension and its tolerance.

7

FIGURE 1-3 [RFS Concept]

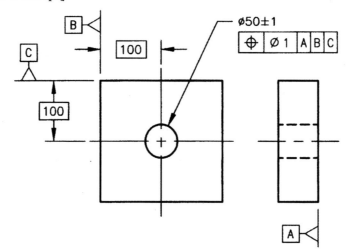

SKIP

**All Boundaries are Non-Constant
in an RFS Call Out** [FIGURE 1-3]

HOLE

Size	Geo. Tol.		Inner Boundary	
Ø49	- 1	=	Ø48	→ worst case
Ø50	- 1	=	Ø49	
Ø51	- 1	=	Ø50	

HOLE

Size	Geo. Tol.		Outer Boundary	
Ø49	+ 1	=	Ø50	
Ø50	+ 1	=	Ø51	
Ø51	+ 1	=	Ø52	→ may be worst case

FIGURE 1-4-Worst Case When the Derived Median Line is Out-of-Straightness

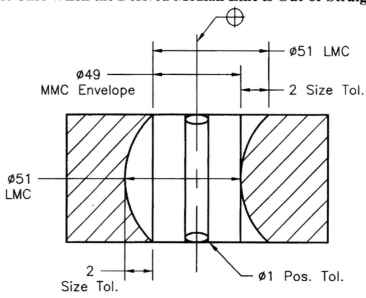

Ø49 + 2 + 2 + 1 = Ø54 worst case outer boundary

Only if the hole has a significant depth might this derived median line curvature (out-of-straightness) be a major factor. For very thin parts, such as sheet metal, it is probably not of concern in these analyses. In fact, many would contend that a banana-shaped hole is not likely to

HOW DO YOU FIND IT?

occur on most products. Therefore, for the analysis in this book, axially out-of-straight holes will not be mentioned.

Creating an Equal Bilateral Toleranced Dimension from Inner and Outer Boundaries
[No Bananas - without the Out-of Axial Straightness Consideration - FIGURE 1-4]

```
  Outer Boundary Hole  =    Ø52
+ Inner Boundary Hole  =    Ø48
        Sum                 Ø100

  Outer Boundary Hole  =    Ø52
+ Inner Boundary Hole  =    Ø48
        Difference     =    Ø4
```

So,

$$\frac{\varnothing 100}{2} = \varnothing 50 \quad \& \quad \frac{4}{2} = 2$$

Therefore, expressed as an equal bilateral toleranced dimension = Ø50±2.

FIGURE 1-5 [Shafts-MMC Concept]

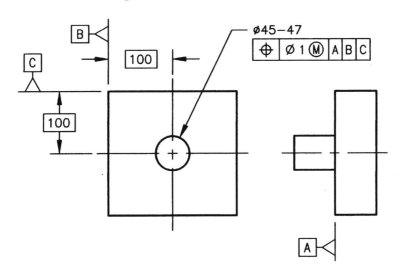

Constant vs. Non-Constant Boundaries [FIGURE 1-5]

SHAFT

Size	Geo. Tol.		Virtual Condition	
Ø47	+ 1	=	Ø48)	→ worst case outer boundary
Ø46	+ 2	=	Ø48)	
Ø45	+ 3	=	Ø48)	

SHAFT

Size	Geo. Tol.		Resultant Condition	
Ø47	− 1	=	Ø46	
Ø46	− 2	=	Ø44	
Ø45	− 3	=	Ø42	→ worst case inner boundary

9

Creating an Equal Bilateral Toleranced Dimension from Virtual and Resultant Conditions [MMC Concept Shafts - FIGURE 1-5]

```
    Resultant Condition Shaft    =   Ø42
 +  Virtual Condition Shaft      =   Ø48
           Sum                   =   Ø90

    Virtual Condition Shaft      =   Ø48
 -  Resultant Condition Shaft    =   Ø42
           Difference            =   Ø6
```

So;

$$\frac{Ø90}{2} = Ø45 \quad \& \quad \frac{6}{2} = 3$$

Therefore, Ø45±3 is an equal bilateral expression of the dimension and its tolerance.

FIGURE 1-6

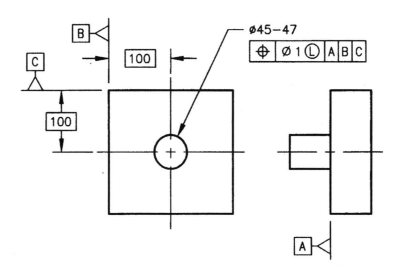

Constant vs. Non-Constant "Worst Case" Boundaries [FIGURE 1-6]

SHAFT

Size	Geo. Tol.	Resultant Condition
Ø45	+ 1	= 46
Ø46	+ 2	= 48
Ø47	+ 3	= 50 → worst case outer boundary

SHAFT

Size	Geo. Tol.	Virtual Condition
Ø45	- 1	= 44) → worst case inner boundary
Ø46	- 2	= 44)
Ø47	- 3	= 44)

SKIP

Creating an Equal Bilateral Toleranced Dimension from Virtual and Resultant Conditions [FIGURE 1-6]

```
  Resultant Condition Shaft   =   Ø50
+ Virtual Condition Shaft     =   Ø44
         Sum                      Ø94

  Resultant Condition Shaft   =   Ø50
- Virtual Condition Shaft     =   Ø44
         Difference           =   Ø6
```

So,

$$\frac{\varnothing 94}{2} = \varnothing 47 \quad \& \quad \frac{6}{2} = 3$$

SKIP

Therefore, expressed as an equal bilateral toleranced dimension = Ø47±3.

FIGURE 1-7

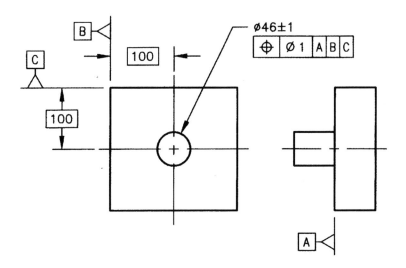

SHAFT

Size	Geo. Tol.		Outer Boundary
Ø45	+ 1	=	Ø46
Ø46	+ 1	=	Ø47
Ø47	+ 1	=	Ø48 → worst case

SHAFT

Size	Geo. Tol.		Inner Boundary
Ø45	- 1	=	Ø44 → worst case (maybe)
Ø46	- 1	=	Ø45
Ø47	- 1	=	Ø46

Creating an Equal Bilateral Toleranced Dimension from Inner and Outer Boundaries

```
  Outer Boundary Shaft    =   Ø48
+ Inner Boundary Shaft    =   Ø44
        Sum                   Ø92

  Outer Boundary Shaft    =   Ø48
- Inner Boundary Shaft    =   Ø44
        Difference        =   Ø4
```

So,

$$\frac{Ø92}{2} = Ø46 \quad \& \quad \frac{4}{2} = 2$$

Therefore, expressed as an equal bilateral toleranced dimension = Ø46±2.

The Ø44mm is identified only as *maybe* the worst case inner boundary. Regardless of feature size conditions have an oddity that must be considered in order to determine the absolute worst case inner boundary.

FIGURE 1-8

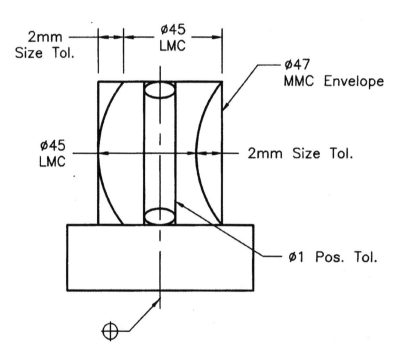

Ø47-2-2-1 = Ø42 worst case inner boundary
[with the derived median line out-of-straightness considered]

Only if the shaft has a significant length might this derived median line out-of-straightness be a major factor. For very short shafts, it is probably not of concern in the analysis. The analyses in this book will not consider this problem, but if your products run the risk of banana-shaped shafts, you may wish to use the above illustration to calculate the worst case inner boundary of your shafts.

CHAPTER 1
EXERCISES

Exercise 1-1

Please convert the following dimensions to equal bilateral dimensions.

1) $\varnothing 100^{+3}_{-1}$

2) $\varnothing 150-155$

3) 200^{+2}_{0}

4) $300^{+0.56}_{-0.43}$

5) $250.06-266.08$

6) $\varnothing 30^{\ 0}_{-0.47}$

7) $500^{+0.26}_{-0.37}$

8) $25.02-25.84$

9) $\varnothing 8.66-8.90$

10) $\varnothing 4.21^{+0.25}_{-0.36}$

Exercise 1-2

Please convert the following specifications to equal bilateral toleranced dimensions generated by their inner and outer boundaries. RFS boundaries have been calculated to reflect holes and shafts that are axially straight and tabs and slots that have flat center planes. So, for this set of problems, dismiss any possibility of out-of-straightness of the derived median lines and out-of-flatness of the derived median planes.

#	Feature & Control	Outer Boundary	Inner Boundary
1	Ø20±2 Hole; ⊕ Ø2 Ⓜ A B C	Ø22	Ø16
2	Ø20±2 Hole; ⊕ Ø2 Ⓛ A B C	Ø24	Ø18
3	Ø20±2 Hole; ⊕ Ø2 A B C	Ø24	Ø16
4	Ø20 +3/−2 Shaft; ⊕ Ø0.5 Ⓜ A B C	Ø23.5	Ø18
5	Ø15–16 Shaft; ⊕ Ø0.4 Ⓛ A B C	Ø16	Ø14.6
6	Ø30 +0.4/−0.2 Shaft; ⊕ Ø0.3 A B C	Ø30.7	Ø29.5
7	28.5 +0.2/−0.1 Slot; ⊕ 0.8 Ⓜ X Y Z	28.7	27.6
8	23.08–24.04 Tab; ⊕ 0.5 Ⓜ T E C	24.54	23.08
9	75±2 Slot; ⊕ 1 Ⓛ C A B	78	73
10	70±0.5 Tab; ⊕ 0.5 Ⓛ D E F	70.5	69
11	12–13 Slot; ⊕ 0.3 G A F	13.3	11.7
12	9.5–10.6 Tab; ⊕ 1.2 H J K	11.8	8.3

Equal bilateral toleranced dimensions:

1) Ø19 ± 3
2) Ø21 ± 3
3) Ø20 ± 4
4) Ø20.75 ± 2.75
5) Ø15.3 ± 0.7
6) Ø30.1 ± 0.6
7) 28.15 ± 0.55
8) 23.81 ± 0.73
9) 75.5 ± 2.5
10) 69.75 ± 0.75
11) 12.5 ± 0.8
12) 10.05 ± 1.75

Chapter 2

TOLERANCE STACK-UP ANALYSIS FOR A BOX ASSEMBLY

• Lesson Objectives:

In Chapter 2, to determine minimum and maximum gaps for a simple eleven-part assembly, you will learn to:
- Perform the calculations.
- Create a loop analysis.
- Create a numbers chart.

Chapter 2
Box Assembly

Step 1:

With this simple box stack-up analysis, we begin by converting the cavity labeled as Part #1 to have an equal bilateral toleranced dimension.

FIGURE 2-1a   [Part #1-The Box]

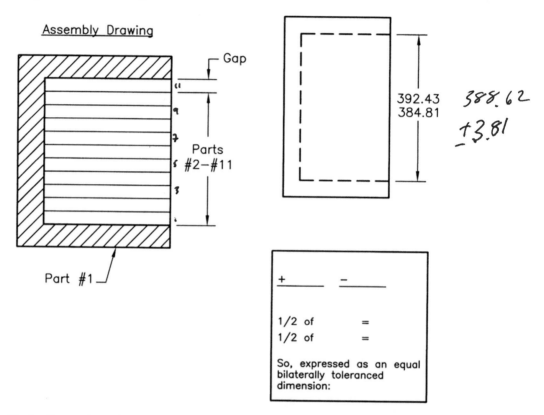

Handwritten annotations: 388.62, ±3.81

Since it has a limit dimension, the upper and lower limits are added and divided by 2 to find the average or mean dimension.

So, 392.43
    +384.81
     777.24   and   777.24 ÷ 2 = 388.62

Then the tolerance is found by subtracting the same limits and dividing by 2 to convert the tolerance to be equal bilateral.

So, 392.43
    - 384.81
       7.62   and   7.62 ÷ 2 = 3.81

Now we know that the limit dimension of 384.81-392.43 expressed as an equal bilaterally toleranced dimension is 388.62±3.81.

FIGURE 2-1b [Parts #2 - #11 = The Plates]

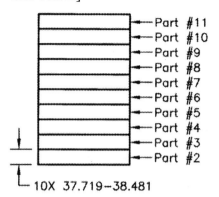

10X 37.719-38.481

```
X 10            X 10            +           −


1/2 of                  =
1/2 of                  =

So, expressed as an equal bilaterally
toleranced dimension:          ±
─────────────────────────────────────
or:
                                /2 =
+           −                   /2 =


So:         ±           ─────10 times
```

The same process is performed on the plates labeled Parts #2 through #11. Since they all have the same dimensions (and the same tolerances), these parts may be done as a group by multiplying their limits by 10 and using them as though all 10 parts were one part, or they may be done individually and added into the minimum and maximum gap loop analysis numbers chart separately.

If these are done as a group, the maximum material condition (MMC) of 38.481 for each is multiplied by 10 (because there are 10 parts) and found to equal 384.81. Then the least material condition (LMC) of 37.719 for each is multiplied by 10 and found to be 377.19.

    MMC   38.481 x 10 = 384.81
    LMC   37.719 x 10 = 377.19

These collective MMC and LMC dimensions are then added.
    So, 384.81
         + 377.19
          762.00

Then to find the mean dimension, the 762 is divided by 2 and found to equal a mean of 381.
    762 ÷ 2 = 381

The total tolerance is calculated by subtracting 377.19 from 384.81 which equals 7.62. This difference is then divided by 2 to equal 3.81.

So, 384.81
  - 377.19
    7.62   and 7.62 ÷ 2 = 3.81

Now we can express the dimension of all 10 plates (Parts #2-#11) as 381±3.81. If this had been done individually, each part (of Parts #2-#11) would have been:

  38.481   34.481
+ 37.719   and - 37.719
  76.2   0.762

and $\frac{76.2}{2} = 38.1$ and $\frac{0.762}{2} = 0.381$

So, Parts #2-#11 would each have been listed as having an equal bilaterally toleranced dimension of 38.1±0.381. This number would have been charted 10 times with the same result as charting it as 381±3.81 one time.

FIGURE 2-2

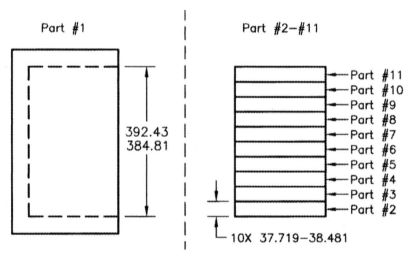

Step 2:

FIGURE 2-3 [Loop Analysis Diagram]

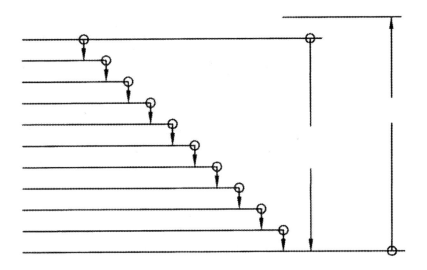

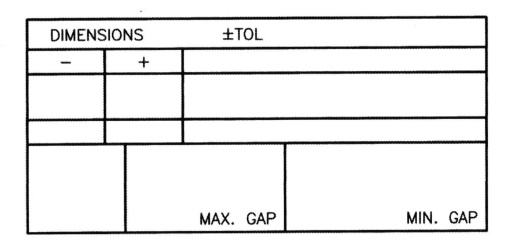

The loop analysis diagram used for this figure begins by showing the gap to be calculated at the top. It begins at Part #11 (a plate) and progressing downward constantly through material until it reaches the last plate at the bottom of the assembly (Plate #2). The sum of all these negative mean dimensions, which run from top to bottom, is 381 and has a total tolerance of plus or minus 3.81.

The loop then progresses up through the cavity (Part #1). This portion of the loop or circuit progresses from bottom to top and is, therefore, designated as positive. The logic of this is simple.

Material is negative (detracting from airspace) and the cavity itself, which has a lack of material, is positive (adding to airspace). The mean dimension of the cavity is 388.62 and has a total tolerance of plus or minus 3.81 (±3.81).

So in the numbers chart, we add the mean dimensions: a positive 388.62 plus a negative 381.00 equals a positive 7.62.

So, +388.62
+ -381.00
7.62

Had this number been a negative sum, it would have proven that even the mean sizes of parts, if produced, would interfere with each other. Since this sum is positive (+7.62), we can proceed.

Step 3:
The next step is to add the charted plus or minus tolerances.

3.81
+3.81
7.62   Tol sum

We are now ready to calculate the minimum and maximum gaps (airspace or interference) by taking the positive mean dimensional difference of 7.62 and adding the sum of the plus and minus tolerances of 7.62 to get the maximum gap (airspace) or subtracting to get the minimum gap (airspace).

So, +7.62 = mean dimensions difference (between positives and negatives)
   + +7.62 = sum of ± tolerances
     15.24 = maximum gap

and +7.62 = mean dimensions difference (between positives and negatives)
   - +7.62 = sum of ± tolerances
        0 = minimum gap

For situations such as this, it may be easier to simply calculate the MMC of the cavity and the collective MMC's of the plates and subtract them to get the minimum gap.

For example: MMC cavity = 384.81

38.481   and   384.81
 x10   - 384.81
384.81=MMC of plates   0=minimum gap

Then calculate the LMC of the cavity and the collective LMC's of the plates and subtract them to get the maximum gap.

For example: LMC cavity = 392.43

37.719   and   392.43
 x10   - 377.19
377.19=LMC of plates   15.24=maximum gap

FIGURE 2-4

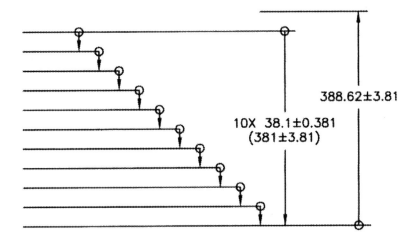

DIMENSIONS		±TOL
−	+	
381		3.81
	388.62	3.81
381	388.62	7.62 TOTALS
388.62 −381.00 ――― 7.62	7.62 +7.62 ――― 15.24=MAX GAP	7.62 −7.62 ――― 0=MIN GAP (zero interference)

Possible solutions to interference problems, should ever they occur:
1. Increase MMC size of cavity.
2. Decrease MMC size of plates.
3. Decrease tolerance on cavity (effectively increasing the MMC).
4. Decrease the tolerances on the plates (effectively decreasing their collective MMC).

CHAPTER 2
EXERCISES

Exercise 2-1, Worksheet #1

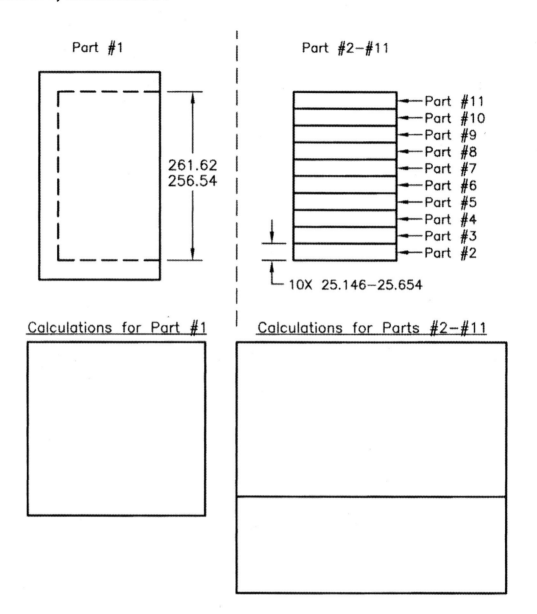

Exercise 2-1, Worksheet #2

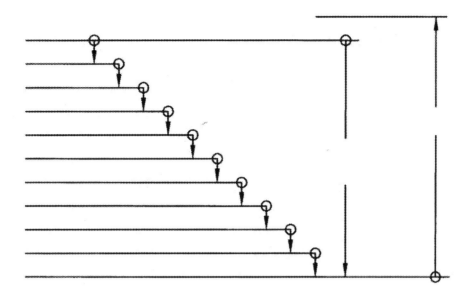

DIMENSIONS		±TOL	
		TOTALS	
Differences in dimensions	MAX GAP		MIN GAP

Chapter 3

TOLERANCE STACK-UP ANALYSIS FOR FEATURE OF SIZE

• Lesson Objectives:

In Chapter 3, you will:
- Perform a loop analysis in both the horizontal and vertical directions to determine MIN and MAX GAPS.
- Determine the proper start and end points of the stack-up analysis.
- Graph the numbers calculated into a loop or circuit diagram.

Chapter 3
Features of Size

FIGURE 3-1 [Tolerance Stack-Up Analysis for Features of Size]

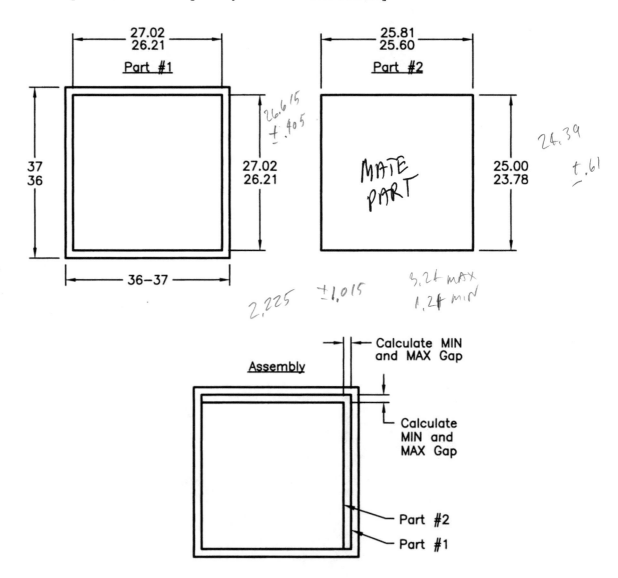

FIGURE 3-2 [Calculations for Tolerance Stack-Up for Features of Size]

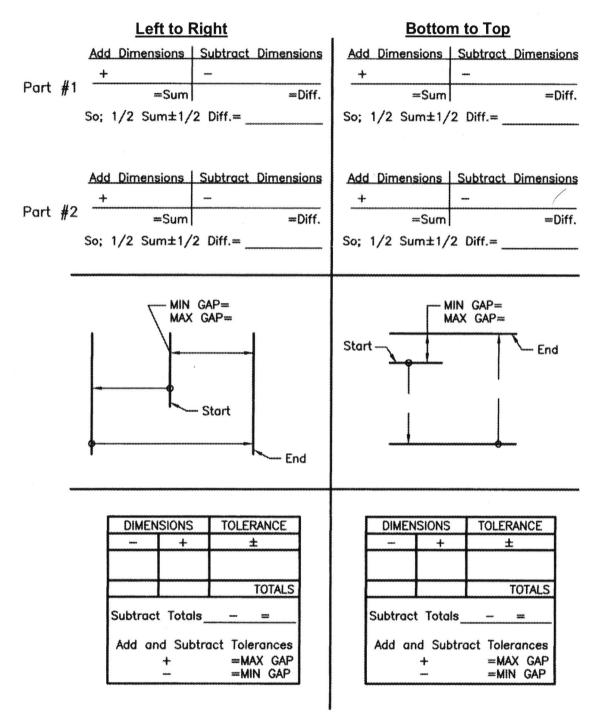

For **this assembly of features of size**, we begin by converting all limit dimensions to equal bilaterally toleranced dimensions. In the horizontal direction, the MMC and LMC of the cavity are converted by adding them and dividing by 2, then subtracting them and dividing by 2.

For example:

```
      27.02 = LMC cavity                            27.02
    + 26.21 = MMC cavity     and                  - 26.21
      53.23 = total                                 0.81 = difference
```
then $\frac{53.23}{2} = 26.615$ and $\frac{0.81}{2} = 0.405$

So, expressed as an equal bilaterally toleranced dimension, it is 26.615±0.405.

The **mating male feature** is treated the same, the mean dimension and plus and minus tolerance calculated as follows:
```
      25.81 = MMC shaft (male feature of size)
    + 25.60 = LMC shaft (male feature of size)     and    51.41/2 = 25.705 mean dimension
      51.41
```
Then: 25.81
   - 25.60   and $\frac{0.21}{2} = 0.105$ plus and minus tolerance
      0.21

So, expressed as an equal bilaterally toleranced dimension, it is 25.705±0.105.

TO BEGIN, the loop analysis is graphed from the left to the right of the gap. So, the route begins at the right of the block and goes left through the material of the male feature as a negative (-) 25.705. Again, in this analysis, the male part is negative in that it steals airspace. Then the loop reverses, going from left to right through the cavity (airspace) or the female graphed as a positive (+) 26.615.

The mean dimensions are charted in this way and added in the numerical charting. In this case:
```
    +26.615 = mean dimension of the cavity (female)
  + -25.705 = mean dimension of the material (male)
    +0.910  = difference between the mean dimensions
```

The plus and minus tolerances are totaled. So,
```
      0.105 = the plus and minus tolerance on the male
    + 0.405 = the plus and minus tolerance on the female
      0.510 = total plus and minus tolerance
```

To **calculate the maximum gap**, the 0.910 difference between the mean dimensions is added to the 0.510 total of the plus and minus tolerances.
So,   0.910
      + 0.510
        1.42 MAX gap

To **calculate the minimum gap**, the 0.510 total of the plus and minus tolerances is subtracted from the 0.910 difference between the mean dimensions.
So,   0.910
       - 0.510
         0.4 MIN gap

In the vertical direction, the MMC and the LMC of the cavity are converted by adding them and dividing by 2, then subtracting them and dividing by 2.

For example:

      27.02 = LMC cavity   27.02
+ 26.21 = MMC cavity   and   - 26.21
  53.23 = total   0.81 = difference

Then: $\frac{53.23}{2} = 26.615$ and $\frac{0.81}{2} = 0.405$

So, expressed as an equal bilaterally toleranced dimension, it is 26.615±0.405.

The mating male feature is treated the same. The mean dimension and the plus and minus tolerance calculated as follows:

     25.00 = MMC shaft (male feature of size)
+ 23.78 = LMC shaft (male feature of size)   and $\frac{48.78}{2} = 24.39$ mean dimension
     48.78

Then: 25.00
    - 23.78   and $\frac{1.22}{2} = 0.61$ plus and minus tolerance
      1.22

So, expressed as an equal bilateral toleranced dimension, it is 24.39±0.61.

TO BEGIN, the loop analysis is graphed from the bottom of the gap to the top of the gap. This means that the route starts at the top of the block and goes down through the material of the male feature as a negative (-) 24.39. Again, in this analysis, the male part is negative in that it steals airspace. Then the loop reverses, going from bottom to top through the cavity (airspace) or the female graphed as a positive (+) 26.615.

The mean dimensions are charted in this way and added in the numerical charting. In this case:

    +26.615 = mean dimension of the cavity (female)
+ -24.390 = mean dimension of the material (male)
     +2.225 = difference between the mean dimensions

The plus and minus tolerances are totaled. So,

    0.610 = the plus and minus tolerance on the male
+ 0.405 = the plus and minus tolerance on the female
    1.015 = total plus and minus tolerance

To **calculate the maximum gap**, the 2.225 difference between the mean dimensions is added to the 1.015 total of the plus and minus tolerances.

So,   2.225
      + 1.015
       3.24 MAX GAP

To **calculate the minimum gap**, the 1.015 total of the plus and minus tolerances is subtracted from the 2.225 difference between the mean dimensions.

So,   2.225
      - 1.015
       1.21 MIN GAP

As in the example shown in Chapter 2, calculating these minimum and maximum gaps may actually be easier by subtracting the MMC's for the minimum gaps and subtracting the LMC's for the maximum gaps. For example:

Horizontal

    26.21 = MMC cavity (Part #1)
  - 25.81 = MMC male (Part #2)
     0.4 = MIN GAP

Vertical

    26.21 = MMC cavity (Part #1)
  - 25.00 = MMC male (Part #2)
     1.21 = MIN GAP

Horizontal

    27.02 = LMC cavity (Part #1)
  -25.60 = LMC male (Part #2)
     1.42 = MAX GAP

Vertical

    27.02 = LMC cavity
  - 23.78 = LMC male
     3.24 = MAX GAP

FIGURE 3-3 [Calculations for Tolerance Stack-Up for Features of Size]

Left to Right

Part #1
```
   27.02      27.02
  +26.21     -26.21
   53.23      0.81 = ±0.405
```
So; 1/2 of 53.23 = 26.615±0.405

Part #2
```
   25.81      25.81
  +25.60     -25.60
   51.41      0.21 = ±0.105
```
So; 1/2 of 51.41 = 25.705±0.105

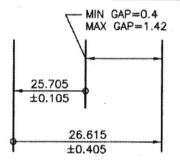

MIN GAP = 0.4
MAX GAP = 1.42

DIMENSIONS		TOLERANCE
−	+	±
25.705		0.105
	26.615	0.405
25.705	26.615	0.510 TOTALS

    26.615
  − 25.705
     0.910

0.91 + 0.51 = 1.42 MAX GAP
0.91 − 0.51 = 0.40 MIN GAP

Bottom to Top

Part #1
```
   27.02      27.02
  +26.21     -26.21
   53.23      0.81 = ±0.405
```
So; 1/2 of 53.23 = 26.615±0.405

Part #2
```
   25.00      25.00
  +23.78     -23.78
   48.78      1.22 = ±0.61
```
So; 1/2 of 48.78 = 24.39±0.61

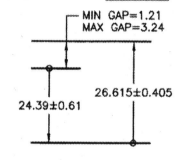

MIN GAP = 1.21
MAX GAP = 3.24

DIMENSIONS		TOLERANCE
−	+	±
24.390		0.610
	26.615	0.405
24.390	26.615	1.015 TOTALS

    26.615
  − 24.390
     2.225

2.225 + 1.015 = 3.24 MAX GAP
2.225 − 1.015 = 1.21 MIN GAP

CHAPTER 3
EXERCISES

Exercise 3-1

[Tolerance Stack-Up Analysis for Features of Size]

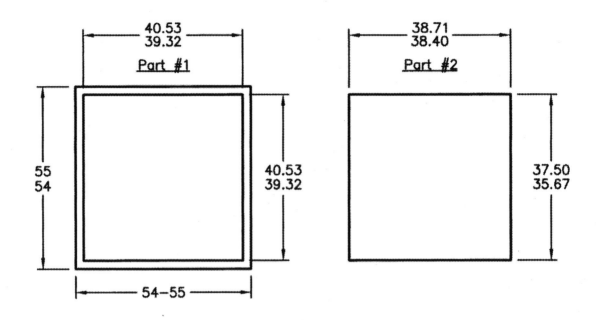

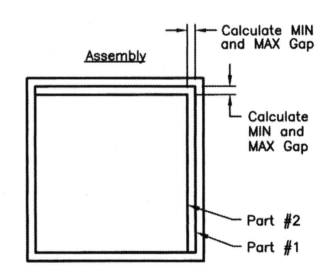

$$\begin{array}{cc} 40.53 & 40.53 \\ +39.32 & -39.32 \\ \hline 2\div\;79.85 & 1.210 \\ & \div\;2 \\ \hline \boxed{39.925 \pm .605} \end{array}$$

$$\begin{array}{cc} 37.50 & 37.50 \\ +35.67 & -35.67 \\ \hline 73.17 & 1.83 \\ \div\;2 & \div\;2 \\ \hline \boxed{36.585 \pm .915} \end{array}$$

Exercise 3-1, Worksheet #1

Left to Right

Part #1

Add Dimensions	Subtract Dimensions
+	−
=Sum	=Diff.

So; 1/2 Sum ± 1/2 Diff. = _____

Part #2

Add Dimensions	Subtract Dimensions
+	−
=Sum	=Diff.

So; 1/2 Sum ± 1/2 Diff. = _____

Bottom to Top

Part #1

Add Dimensions	Subtract Dimensions
+	−
=Sum	=Diff.

So; 1/2 Sum ± 1/2 Diff. = _____

Part #2

Add Dimensions	Subtract Dimensions
+	−
=Sum	=Diff.

So; 1/2 Sum ± 1/2 Diff. = _____

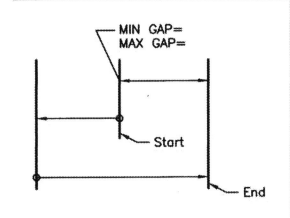

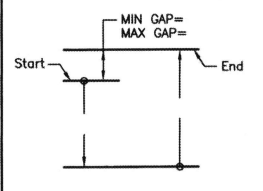

DIMENSIONS		TOLERANCE
−	+	±
		TOTALS

Subtract Totals ____ − ____ = ____

Add and Subtract Tolerances
+ ____ =MAX GAP
− ____ =MIN GAP

DIMENSIONS		TOLERANCE
−	+	±
		TOTALS

Subtract Totals ____ − ____ = ____

Add and Subtract Tolerances
+ ____ =MAX GAP
− ____ =MIN GAP

Chapter 4

TOLERANCE STACK-UP ANALYSIS FOR ASSEMBLIES WITH PLUS AND MINUS TOLERANCING

• Lesson Objectives:

In Chapter 4, you will:
- Calculate the airspace and interferences for a plus and minus toleranced assembly.
- Perform multiple loop analyses on an assembly.

Chapter 4
Assemblies with Plus and Minus Tolerances

FIGURE 4-1

33.325 ± 1.575

25.805
± .405

24.995
± .405

26.21   26.21
25.40   25.40

20.625
± .785

15.88
± .790

SEE SH 42

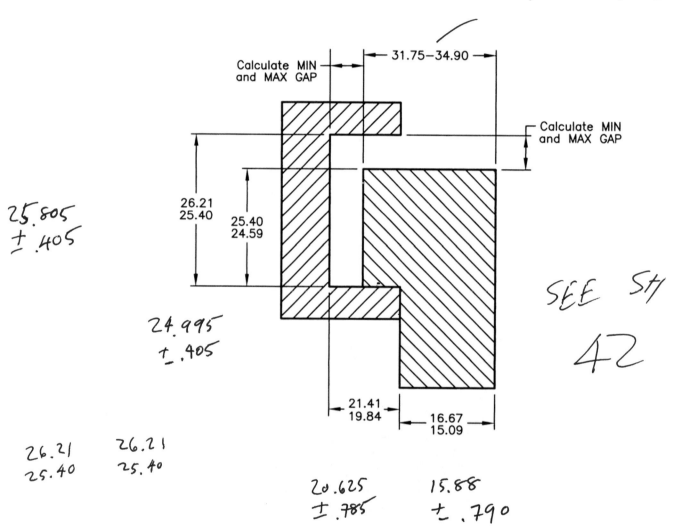

FIGURE 4-2

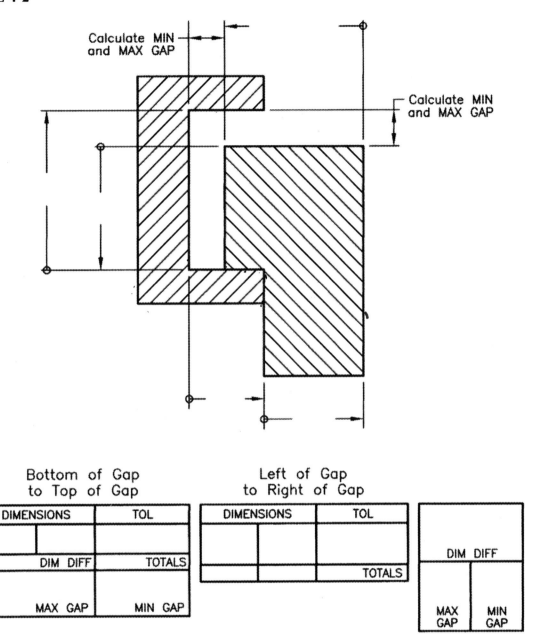

This assembly is also one that relies on plus and minus tolerances or limit dimensions to tolerance the mating features. So, again, we begin by deciding which gap we wish to calculate. Then we convert all pertinent dimensions and their tolerances to equal bilaterally toleranced dimensions. In this assembly, we will calculate MIN and MAX GAPS in both the vertical and the horizontal directions, so we may as well begin by converting all of the dimensions to have equal bilaterally tolerances.

To find the mean dimensions, we can take each set of upper and lower limit dimensions, add them and then divide by 2. To find their equal bilateral tolerances, we subtract each set and then divide by 2.

Horizontal Direction:

In the horizontal direction, we perform the following calculations:

$$34.90 = \text{MMC male}$$
$$+\,31.75 = \text{LMC male}$$
$$66.65 = \text{sum}$$

and

$$34.90 = \text{MMC male}$$
$$-\,31.75 = \text{LMC male}$$
$$3.15 = \text{difference}$$

then $\dfrac{66.65}{2} = 33.325$ and $\dfrac{3.15}{2} = 1.575$

Therefore, expressed as an equal bilateral toleranced dimension, it is 33.325±1.575.

The next dimension to be converted in the horizontal direction is 19.84 - 21.41. So,

$$21.41 = \text{LMC air}$$
$$+\,19.84 = \text{MMC air}$$
$$41.25 = \text{sum}$$

and

$$21.41 = \text{LMC air}$$
$$-\,19.84 = \text{MMC air}$$
$$1.57 = \text{difference}$$

then $\dfrac{41.25}{2} = 20.625$ and $\dfrac{1.57}{2} = 0.785$

Therefore, expressed as an equal bilateral toleranced dimension, it is 20.625±0.785.

The next dimension to be converted in the horizontal direction is 15.09 - 16.67. So,

$$16.67 = \text{MMC male}$$
$$+\,15.09 = \text{LMC male}$$
$$31.76 = \text{sum}$$

and

$$16.67$$
$$-\,15.09$$
$$1.58 = \text{difference}$$

then $\dfrac{31.76}{2} = 15.88$ and $\dfrac{1.58}{2} = 0.79$

Therefore, expressed as an equal bilaterally toleranced dimension, it is 15.88±0.79.

Vertical Direction:

In the vertical direction, one of the pertinent dimensions is 25.40 - 26.21. So,

$$26.21 = \text{air LMC}$$
$$+\,25.40 = \text{air MMC}$$
$$51.61 = \text{sum}$$

and

$$26.21$$
$$-\,25.40$$
$$0.81 = \text{difference}$$

then $\dfrac{51.61}{2} = 25.805$ and $\dfrac{0.81}{2} = 0.405$

So, as an equal bilaterally toleranced dimension, it is 25.805±0.405.

The other pertinent dimension in the vertical direction is 24.59 - 25.40. So,

$$25.40 = \text{MMC (material) male}$$
$$+\,24.59 = \text{LMC (material) male}$$
$$49.99 = \text{sum}$$

and

$$25.40$$
$$-\,24.59$$
$$0.81 = \text{difference}$$

then $\dfrac{49.99}{2} = 24.995$ and $\dfrac{0.81}{2} = 0.405$

So, expressed as an equal bilateral, it is 24.995±0.405.

FIGURE 4-3 [Tolerance Stack-Up Calculations for Plus and Minus Tolerances Left to Right]

$$
\begin{array}{cccccc}
34.90 & 34.90 & 21.41 & 21.41 & 16.67 & 16.67 \\
+\,31.75 & -\,31.75 & +\,19.84 & -\,19.84 & +\,15.09 & -\,15.09 \\
\hline
66.65 & 3.15 & 41.25 & 1.57 & 31.76 & 1.58
\end{array}
$$

66.65/2=33.325   41.25/2=20.625   31.76/2=15.88

3.15/2=1.575   1.57/2=0.785   1.58/2=0.79

33.325 ± 1.575   20.625 ± 0.785   15.88 ± 0.79

$$
\begin{array}{cccc}
26.21 & 26.21 & 25.40 & 25.40 \\
+\,25.40 & -\,25.40 & +\,24.59 & -\,24.59 \\
\hline
51.61 & 0.81 & 49.99 & 0.81
\end{array}
$$

51.61/2=25.805   49.99/2=24.995

0.81/2=0.405   0.81/2=0.405

25.805 ± 0.405   24.995 ± 0.405

Bottom to Top of the Gap
To begin, the loop analysis is graphed from the bottom of the gap to the top of the gap. So, we begin at the top of the block and go down through the material of the male feature as a negative number of -24.995. Again, in this analysis, the male part is negative in that it steals airspace. Then the loop reverses to going from bottom to top through the cavity (airspace) or female, which is graphed as a positive number of +25.805.

The mean dimensions are charted in this way and added in the numerical charting. In this case,
    +25.805 = mean dimension of the cavity (female)
 + -24.995 = mean dimension of the material (male)
     +0.810 = difference between the mean dimensions

The plus and minus tolerances are totaled. So,
    0.405 = the plus and minus tolerance on the male
 + 0.405 = the plus and minus tolerance on the female
    0.81 = total plus and minus tolerance

To **calculate the maximum gap**, the 0.81 difference between the mean dimensions is added to the 0.81 total of the plus and minus tolerances. So,
    0.81
 + 0.81
    1.62 = MAX GAP

To **calculate the minimum gap**, the 0.81 total of the plus and minus tolerances is subtracted from the 0.81 difference between the mean dimensions. So,
    0.81
 - 0.81
    0 = MIN GAP

As in the examples shown in Chapters 2 and 3, calculating these minimum and maximum gaps (airspaces) may actually be easier by subtracting the MMC's for the minimum gaps and subtracting the LMC's for the maximum gaps. For example:

                  Vertical   **Vertical**

            25.40 = MMC cavity   26.21 = LMC cavity

         - 25.40 = MMC male   and   - 24.59 = LMC male

            0   = MIN GAP   1.62 = MAX GAP

Left of the Gap to the Right of the Gap

In the horizontal direction, the loop analysis and number charting will begin by designating all numbers proceeding from right to left as negative, and all numbers from left to right as positive. Since the 15.88 dimension is seen as a part of the 33.325 dimension, the 15.88 will be designated as a positive. Like the 20.625 dimension, it can be looked at as adding airspace by reducing the overall material of the 33.325. So the 20.625 is positive and the 15.88 is positive; and, when added together, the sum is 36.505.

        +20.625
  + +15.880
        +36.505

Since the only negative number is 33.325, we can now add it to the positive 36.505, and the sum or positive difference is the mean total. So,

        +36.505
  + -33.325
          3.180 = mean total

To **calculate the maximum gap**, we add the mean total to the total of all the plus and minus tolerances. The tolerance total is:

          0.785 = ± tolerance on 20.625
          0.790 = ± tolerance on 15.88
       + 1.575 = ± tolerance on 33.325
          3.150 = total ± tolerance

So,   3.18 = mean total
              + 3.15 = total of the ± tolerances
                6.33 = MAX GAP

And to calculate the minimum gap, we subtract the same two numbers. So,

           3.18 = mean total
       - 3.15 = total of the ± tolerances
          0.03 = MIN GAP

Logic would lead us to the conclusion that **the minimum gap** could be calculated by using the largest dimension, or the 34.90, on the overall male and the smallest dimension (taking away the smallest amount from the 34.90) on the other male portion, which is 15.09. Then, the other pertinent number for minimum gap is the smallest height of the cavity, which is 19.84. So, these numbers together would be:

34.90 = MMC male #1		19.84 = MMC cavity (lack of material)	
- 15.09 = LMC male #2	and	- 19.81 = MMC material	
19.81 = MMC material		0.03 = MIN GAP	

To **determine the maximum gap**, the calculation should employ the 31.75 LMC of the male feature #1 and the 16.67 MMC of the male feature #2. This would provide the shortest extension of material over the ledge. Then, the 21.41 LMC of the cavity depth would be the deepest, allowing the MAX GAP. So,

31.75 = LMC male #1		21.41 = LMC cavity
- 16.67 = MMC male #2	and	- 15.08 = LMC material
15.08 = LMC material		6.33 = MAX GAP

FIGURE 4-4 [Tolerance Stack-Up Analysis for Plus and Minus Tolerances]

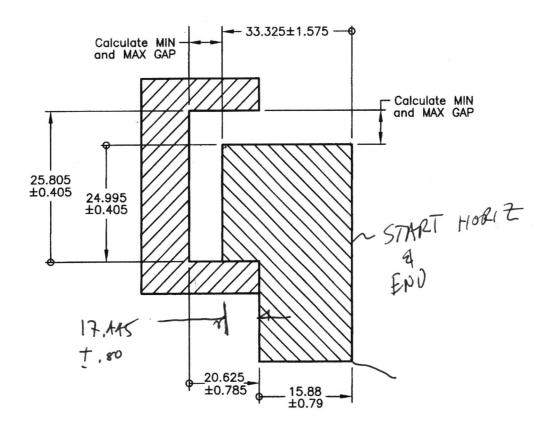

DIMENSIONS		TOL		DIMENSIONS		TOL		36.505 -33.325	
-24.995	+25.805	0.405 0.405		+20.625 +15.880	-33.325	0.785 0.790 1.575		3.180 DIM DIFF	
+0.810 DIM DIFF		0.810 TOTALS		+36.505	-33.325	3.150 TOTALS		3.180 +3.150	3.180 -3.150
0.81 +0.81		0.81 -0.81						6.330 MAX GAP	0.030 MIN GAP
1.62=MAX GAP		0=MIN GAP							

CHAPTER 4
EXERCISES

Exercise 4-1

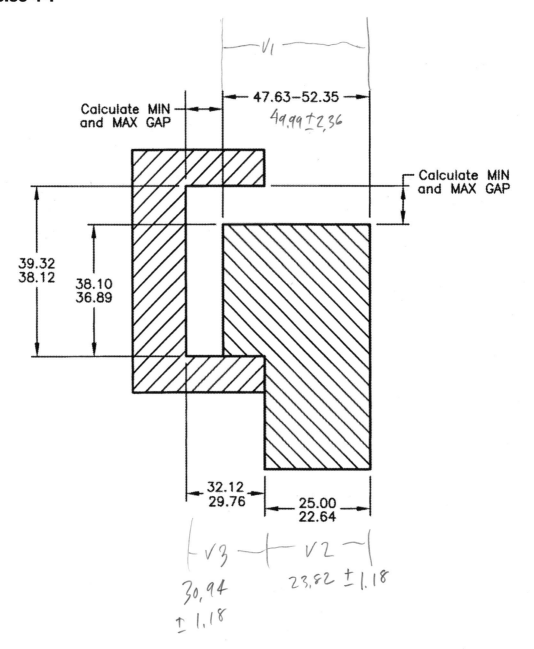

Exercise 4-1, Worksheet #1

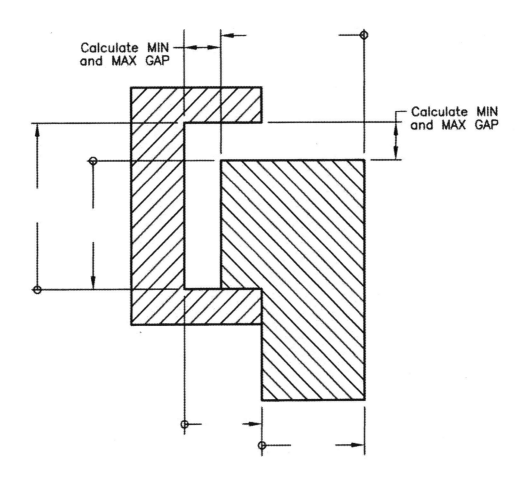

Chapter 5

TOLERANCE STACK-UP ANALYSIS FOR A FLOATING FASTENER ASSEMBLY USING GEOMETRIC TOLERANCES

• Lesson Objectives:

In Chapter 5, for a geometrically-toleranced floating fastener assembly, you will:
- Calculate resultant conditions and virtual conditions, inner boundaries and outer boundaries.
- Determine the mean of the boundaries and conditions.
- Convert all diameters and full widths to mean radii with equal bilateral tolerances.
- Mix widths and diameters in a numbers chart.
- Graph the numbers into a tolerance stack-up diagram.
- Determine all unknown gaps in a seven-part assembly.

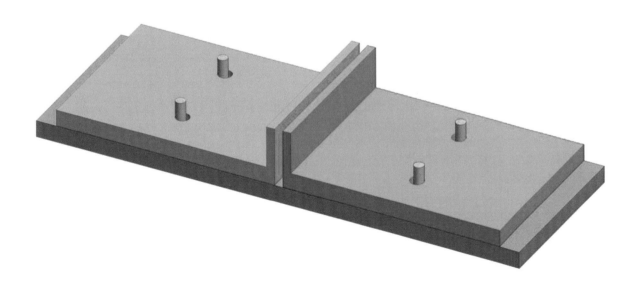

Chapter 5
A Floating Fastener Assembly using Geometric Tolerances

FIGURE 5-1

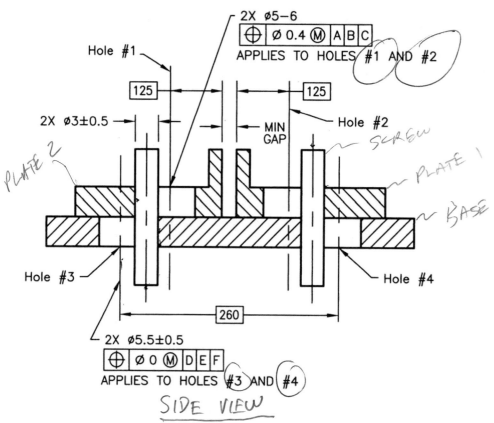

This assembly consists of two identical parts mounted onto a larger plate. Each of smaller parts has two clearance holes that are used to align and bind the parts to clearance holes on the larger plate. The pins shown passing through these holes are a generic representation of screws. The heads of the screws, their threads and the nuts that bind the assembly together have not been shown in this illustration.

The procedure demonstrated here is used for floating fastener assemblies. Since the stack-up problem to be worked would be the same no matter how many holes each part had within the pattern, only those necessary for the analysis are shown in the detail drawings. For a 3-Dimensional look at the assembly please see the first illustration in this chapter.

Handwritten annotations at top:
- RESULTANT (WORST-CASE) (OUTER BOUNDRY) NON CONSTANT
- VIRTUAL (INNER BOUNDRY) CONSTANT

FIGURE 5-2

Resultant Condition of Hole #1= _____
Virtual Condition of Hole #1= _____

Resultant Condition of Hole #2 _____
Virtual Condition of Hole #2 _____

Resultant Condition of Hole #3 _____
Virtual Condition of Hole #3 _____

Resultant Condition of Hole #4 _____
Virtual Condition of Hole #4 _____

   Res. Cond. Hole #1=   Res. Cond. Hole #1=
+ Virt. Cond. Hole #1=_____   - Virt. Cond. Hole #1=_____
        Sum   Difference

Sum and Difference Divided by 2=_____ =Mean Dia. ± Mean Tol.
Mean Dia. ± Mean Tol. Divided by 2=_____ =Mean Radius ± Mean Tol.

   Res. Cond. Hole #2=   Res. Cond. Hole #2=
+ Virt. Cond. Hole #2=_____   - Virt. Cond. Hole #2=_____
        Sum   Difference

Sum and Difference Divided by 2=_____ =Mean Dia. ± Mean Tol.
Mean Dia. ± Mean Tol. Divided by 2=_____ =Mean Radius ± Mean Tol.

   Res. Cond. Hole #3=   Res. Cond. Hole #3=
+ Virt. Cond. Hole #3=_____   - Virt. Cond. Hole #3=_____
        Sum   Difference

Sum and Difference Divided by 2=_____ =Mean Dia. ± Mean Tol.
Mean Dia. ± Mean Tol. Divided by 2=_____ =Mean Radius ± Mean Tol.

   Res. Cond. Hole #4=   Res. Cond. Hole #4=
+ Virt. Cond. Hole #4=_____   - Virt. Cond. Hole #4=_____
        Sum   Difference

Sum and Difference Divided by 2=_____ =Mean Dia. ± Mean Tol.
Mean Dia. ± Mean Tol. Divided by 2=_____ =Mean Radius ± Mean Tol.

In the example shown in FIGURE 5-1, minimum and maximum gaps are calculated for a floating fastener assembly where geometric tolerances have been specified. To work with a geometrically toleranced feature, inner and outer boundaries must be calculated. These worst case inner and outer boundaries are the collective effect of size and geometric tolerance. If the boundaries are constant, they are known as virtual conditions. If they are not constant, each has a worst case that is represented. These worst case boundaries are called resultant conditions. Only the worst case boundaries are applicable to the tolerance stack-up analysis.

To begin, every virtual condition and resultant condition is calculated. The four holes are referenced in their feature control frames at MMC. This means that their inner boundaries will be constant virtual conditions. It also means that their outer boundaries will vary with each size they are produced at. Therefore, the only outer boundary we will calculate will be the largest (the worst case).

Hole #1 and Hole #2 have the same size and geometric tolerances, so their calculation is:

$$
\begin{array}{ll}
6.0 = \text{LMC Hole} & 5.0 = \text{MMC Hole} \\
+1.4 = \text{Geo. Tol. at LMC} \quad \text{and} & -0.4 = \text{Geo. Tol. at MMC} \\
\varnothing 7.4 = \text{Resultant Condition Hole} & \varnothing 4.6 = \text{Virtual Condition Hole} \\
\quad \text{(worst case outer boundary)} & \quad \text{(constant inner boundary)}
\end{array}
$$

Hole #3 and Hole #4 have the same size and geometric tolerances, so their calculation is:

$$
\begin{array}{ll}
6.0 = \text{LMC Hole} & 5.0 = \text{MMC Hole} \\
+1.0 = \text{Geo. Tol. at LMC} \quad \text{and} & -0.0 = \text{Geo. Tol. at MMC} \\
\varnothing 7.0 = \text{Resultant Condition Hole} & \varnothing 5.0 = \text{Virtual Condition Hole} \\
\quad \text{(worst case outer boundary)} & \quad \text{(constant inner boundary)}
\end{array}
$$

Now that the inner and outer boundaries have been calculated and recorded, we must determine the difference between each hole's resultant condition (outer boundary) and virtual condition (inner boundary).

Hole #1
The resultant condition of Hole #1 is Ø7.4, and its virtual condition is Ø4.6.
The difference is Ø2.8.

Hole #2
The resultant condition of Hole #2 is Ø7.4, and its virtual condition is Ø4.6.
The difference is Ø2.8.

Hole #3
The resultant condition of Hole #3 is Ø7, and its virtual condition is Ø5.
The difference is Ø2.

Hole #4
The resultant condition of Hole #4 is Ø7, and its virtual condition is Ø5.
The difference is Ø2.

Each difference is then divided by 2 to arrive at the mean tolerance in diameter form. So, for Holes #1 and #2, we divide Ø2.8 by 2 and get Ø1.4. And for Holes #3 and #4, we divide Ø2 by 2 and get Ø1.

The next step is to add the resultant condition and virtual condition of each hole and divide it by 2 to get the mean diameter. So, for Holes #1 and #2, we add the virtual condition of Ø4.6 to the resultant condition of Ø7.4 to find the sum of Ø12. The Ø12 is then divided by 2 to get the mean diameter of 6. For example:

    Ø7.4 = Resultant Condition
    + Ø4.6 = Virtual Condition   Then, $\frac{Ø12}{2} = Ø6$
    Ø12.0 = Sum

Holes #3 and #4 have resultant conditions of a Ø7 and virtual conditions of a Ø5, which are added together to get a Ø12. The Ø12 is then divided by 2 to get the mean diameter of 6. For example:

    Ø7 = Resultant Condition
    + Ø5 = Virtual Condition   Then, $\frac{Ø12}{2} = Ø6$
    Ø12 = Sum

Now we know the mean diameter and mean tolerance for each hole. Expressed as a diameter with a plus and minus tolerance, they are:
    Hole #1 = Ø6±1.4
    Hole #2 = Ø6±1.4
    Hole #3 = Ø6±1
    Hole #4 = Ø6±1

Since we will calculate our MIN and MAX GAPS using the holes' radii, each of these dimensions and their tolerances are converted to radii by dividing everything by 2 again. So, each hole's mean diameter and its plus and minus tolerance, expressed as a mean radius with a plus and minus tolerance, is:
    Hole #1 = R3±0.7
    Hole #2 = R3±0.7
    Hole #3 = R3±0.5
    Hole #4 = R3±0.5

FIGURE 5-3

Resultant Condition of Hole #1= ⌀6.0 + 1.4=⌀7.4
Virtual Condition of Hole #1=⌀5.0-0.4=⌀4.6

Resultant Condition of Hole #2=⌀6.0 + 1.4=⌀7.4
Virtual Condition of Hole #2=⌀5.0-0.4=⌀4.6

Resultant Condition of Hole #3=⌀6.0 + 1.0=⌀7.0
Virtual Condition of Hole #3=⌀5.0-0=⌀5.0

Resultant Condition of Hole #4=⌀6.0 + 1.0=⌀7.0
Virtual Condition of Hole #4=⌀5.0-0=⌀5.0

Res. Cond. Hole #1= ⌀7.4	Res. Cond. Hole #1=⌀7.4
+ Virt. Cond. Hole #1=⌀4.6	- Virt. Cond. Hole#1=⌀4.6
Sum 12	Difference 2.8

Sum and Difference Divided by 2=⌀6±1.4=Mean Dia. ± Mean Tol.
Mean Dia. ± Mean Tol. Divided by 2=R3±0.7=Mean Radius ± Mean Tol.

Res. Cond. Hole #2= ⌀7.4	Res. Cond. Hole #2=⌀7.4
+ Virt. Cond. Hole #2=⌀4.6	- Virt. Cond. Hole#2=⌀4.6
Sum 12	Difference 2.8

Sum and Difference Divided by 2=⌀6±1.4=Mean Dia. ± Mean Tol.
Mean Dia. ± Mean Tol. Divided by 2=R3±0.7=Mean Radius ± Mean Tol.

Res. Cond. Hole #3= ⌀7	Res. Cond. Hole #3=⌀7
+ Virt. Cond. Hole #3=⌀5	- Virt. Cond.Hole #3=⌀5
Sum 12	Difference 2

Sum and Difference Divided by 2=⌀6±1=Mean Dia. ± Mean Tol.
Mean Dia. ± Mean Tol. Divided by 2=R3±0.5=Mean Radius ± Mean Tol.

Res. Cond. Hole #3= ⌀7	Res. Cond. Hole #3=⌀7
+ Virt. Cond. Hole #3=⌀5	- Virt. Cond.Hole #3=⌀5
Sum 12	Difference 2

Sum and Difference Divided by 2=⌀6±1=Mean Dia. ± Mean Tol.
Mean Dia. ± Mean Tol. Divided by 2=R3±0.5=Mean Radius ± Mean Tol.

FIGURE 5-4

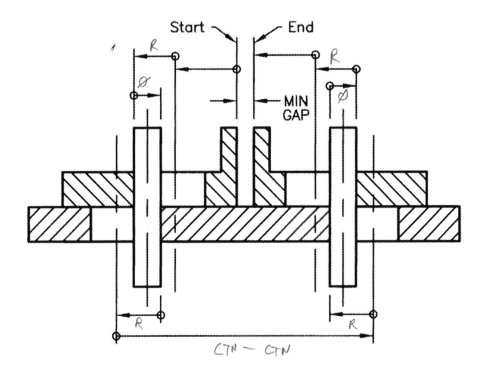

Loop Diagram

Right to Left	Left to Right	±Tol	
−	+		
			Totals
			MIN GAP

FIGURE 5-5 [The loop diagram and its accompanying numbers chart are filled in.]

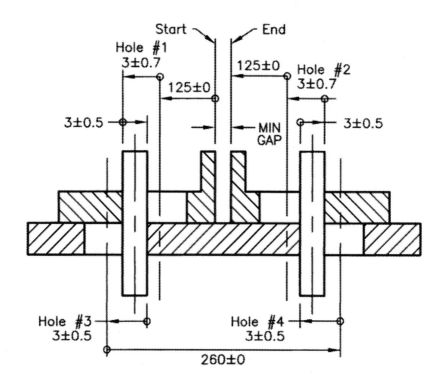

Loop Diagram

Right to Left	Left to Right	±Tol	
−	+		
125		0.0	Basic Dim
3		0.7	Hole #1
	3	0.5	Pin
3		0.5	Hole #3
	260	0.0	Basic Dim
3		0.5	Hole #4
	3	0.5	Pin
3		0.7	Hole #2
125		0.0	Basic Dim
262	266	3.4	Totals

```
266          4.0
-262        -3.4
----        ----
  4          0.6  MIN GAP
```

The loop diagram begins by pushing all parts in such a fashion that would create the minimum gap. As FIGURE 5-5 shows, this has had the effect of trapping the pins between Holes #1 and #3 and also between Holes #2 and #4.

The pin on the left passing through Holes #1 and #3 is trapped by pushing the part that contains Hole #1 to the right (closing the gap between the parts containing Holes #1 and #2). The pin on the right passing through Holes #2 and #4 is trapped by pushing the part that contains Hole #2 to the left (closing the gap between the parts containing Hole #1 and #2).

Since this is a floating fastener situation, the pin on the left simply slides along until it is against the left side of Hole #1 and against the right side of Hole #3. The pin on the right slides until it rests against the right side of Hole #2 and against the left side of Hole #4. The gap being calculated is now at its minimum.

STEP 1:
The loop begins at the wall forming the left side of the gap. It proceeds to the left, designated as negative numbers, through the basic dimension of 125mm to the middle of Hole #1.

STEP 2:
Step 2 proceeds 3mm left (negative) again through the radius of Hole #1 (as calculated with its virtual and resultant condition boundaries). Hole #1 is complete.

STEP 3:
Now we progress back through the diameter of the pin trapped between Holes #1 and #3. This is designated as a positive 3mm (all number routes to the right are designated as positive).

STEP 4:
Now we involve Hole #3 by following the loop from the right edge of the hole to Hole #3's center as a negative 3mm.

STEP 5:
This step takes us left to right (positive) from the middle of Hole #3 to the middle of Hole #4 the basic dimension of 260mm. Hole #3 is now complete.

STEP 6:
Step 6 reverses the route going left (negative) through the 3mm radius of Hole #4. Hole #4 is complete.

STEP 7:
Step 7 reverses the loop again and goes right through the diameter of the last pin a positive 3mm.

STEP 8:
Step 8 begins on the last hole, Hole #2. The route goes from its right edge to its middle (a right to left route) a negative 3mm.

STEP 9:
Step 9 continues left (negative) from the center of Hole #2 to the end of the gap a basic dimension of 125mm.

The negative numbers (right to left routes) in the loop were 125, 3, 3, 3, 3 and 125, for a negative sum of 262. The positive numbers (left to right routes) in the loop were 3, 260 and 3, for a positive sum of 266. The sums added:

```
    +266
+   -262
    +4
```

So a positive 4 is the minimum gap before tolerances are used. The tolerances on the dimensions were:

STEP 1	0.0 =	Basic Dimension of 125
STEP 2	0.7 =	Hole #1
STEP 3	0.5 =	Pin
STEP 4	0.5 =	Hole #3
STEP 5	0.0 =	Basic Dimension of 260
STEP 6	0.5 =	Hole #4
STEP 7	0.5 =	Pin
STEP 8	0.7 =	Hole #2
STEP 9	0.0 =	Basic Dimension of 125

The sum of these tolerances is 3.4. So, the minimum gap is:
+4.0
-3.4
0.6 = MIN GAP

The logic behind this loop route was to proceed from the left edge of the gap through all features having an effect on the MIN GAP size, to the right edge of the gap. The route went left or right as appropriate so as to involve all pertinent features until the loop was complete. The pertinent features were the four holes and the two pins passing through them. Hole radii were used because the pertinent basic dimensions binding the gap to the holes, and the holes to each other, went to the hole centers. The full pin diameters were used because they were sandwiched between (jammed against the opposing sides) the holes. The basic dimensions were used to allow a route from the left side of the MIN GAP to the center of Hole #1, then from the center of Hole #3 to the center of Hole #4, then from the center of Hole #3 to the right side of the MIN GAP.

The route was:
- **Step 1** led from the left side of the MIN GAP to the center of Hole #1 (-125mm).
- **Step 2** led from the center of Hole #1 to the left side of Hole #1 (-3mm).
- **Step 3** led from the left side of the pin to the right side of the Pin (+3mm).
- **Step 4** led from the right side of the pin to the center of Hole #3 (-3mm).
- **Step 5** led from the center of Hole #3 to the center of Hole #4 (+260mm).
- **Step 6** led from the center of Hole #4 to the left side of Hole #4 (-3mm).
- **Step 7** led from the left side of the pin to the right side of the Pin (+3mm).
- **Step 8** led from the right side of Hole #2 to the center of Hole #2 (-3mm).
- **Step 9** led from the center of Hole #2 to the right side of the gap (a right to left route, though) (-125mm).

If you examine the route taken, you can see that, with the parts shoved to create the MIN GAP (given the MIN GAP was a positive number), this is the only logical route to take. It gets the tolerance analyst from the left side of the gap to the right side of the gap by taking the only known route of numbers. Analyzing the positive numerical result and seeing that this route was also the shortest route is a very good indication that the answer is correct (providing there were no math errors). Remember, this is a house of cards. One error of any kind and the house falls.

FIGURE 5-6

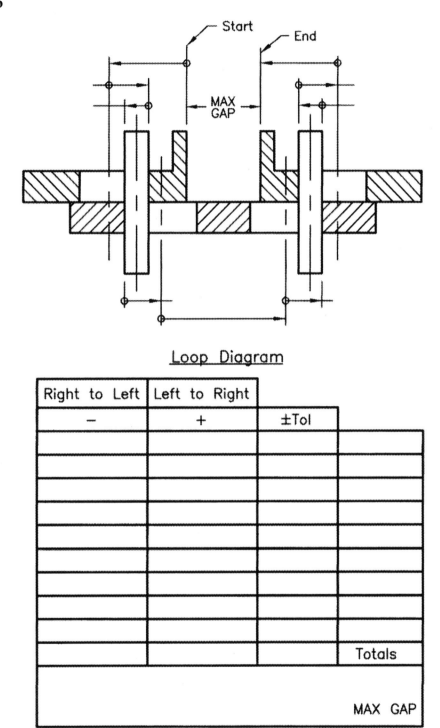

The next loop diagram begins by pushing all parts in such a way as to create the maximum gap. As FIGURE 5-6 shows, this has the effect of trapping the pins between Holes #1 and #3, and also between Holes #2 and #4. The pin on the left, passing through Hole #1 and Hole #3, is trapped by pushing the part that contains Hole #1 to the left, opening the gap between the parts that contain Holes #1 and #2. The pin on the right, passing through Holes #2 and #4 is trapped by pushing the part that contains Hole #2 to the right, opening the gap between the parts containing Holes #1 and #2.

Since this is a floating fastener assembly, the pin on the left simply slides along until it is against the right side of Hole #1 and the left side of Hole #3. The pin on the right slides until it is against the left side of Hole #2 and the right side of Hole #4. The gap being calculated is now at its maximum.

FIGURE 5-7

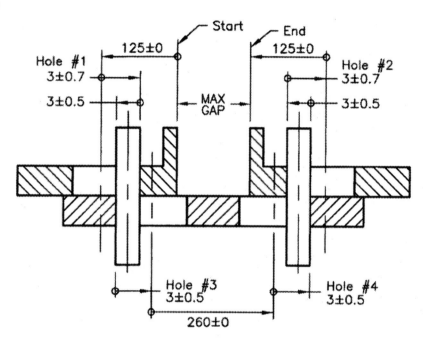

Loop Diagram

Right to Left	Left to Right	±Tol	
−	+		
125		0.0	Basic Dim
	3	0.7	Hole #1
3		0.5	Pin
	3	0.5	Hole #3
	260	0.0	Basic Dim
	3	0.5	Hole #4
3		0.5	Pin
	3	0.7	Hole #2
125		0.0	Basic Dim
256	272	3.4	Totals
272 −256 16		16.0 +3.4 19.4 MAX GAP	

STEP 1:
The loop begins at the wall forming the left side of the gap. It proceeds to the left, designated as negative numbers, through the basic dimension of 125mm to the middle of Hole #1.

STEP 2:
Step 2 proceeds 3mm to the right (positive) through the radius of Hole #1 (as calculated by its virtual and resultant condition boundaries. Hole #1 is now complete.

STEP 3:
Step 3 reverses the loop to the left, back through the diameter of the pin trapped between Holes #1 and #3. This is designated as a negative 3mm (all number routes that proceed from right to left are designated as negative).

STEP 4:
Now we involve Hole #3 by following the loop from the left edge of Hole #3 to the center of Hole #3 (left to right) a positive 3mm.

STEP 5:
This step takes us from the center of Hole #3 to the center of Hole #4 (left to right) a positive 260mm. Hole #3 is now complete.

STEP 6:
Step 6 keeps going right (positive) to the right edge of Hole #4 a positive 3mm.

STEP 7:
Step 7 reverses the loop and goes from the right side of the diameter of the last pin to the left side of the pin a negative 3mm.

STEP 8:
Step 8 begins on the last hole, Hole #2. The route goes from the left side of Hole #2 to the center of Hole #2 a positive 3mm.

STEP 9:
Step 9 reverses the route for the last time, going from right to left the basic dimension a negative 125mm to the end of the gap.

The negative numbers (right to left routes) in the loop were 125, 3, 3 and 125, for a negative total of 256. The positive numbers (left to right routes) in the loop were 3, 3, 260, 3 and 3, for a positive total of 272. The sums added:

$$+272$$
$$+\underline{-256}$$
$$+16$$

So a positive 16 is the maximum gap before tolerances are used. The tolerances on the dimensions were:

STEP 1   0.0 =   Basic Dimension of 125
STEP 2   0.7 =   Hole #1
STEP 3   0.5 =   Pin
STEP 4   0.5 =   Hole #3
STEP 5   0.0 =   Basic Dimension of 260
STEP 6   0.5 =   Hole #4
STEP 7   0.5 =   Pin

STEP 8   0.7 =   Hole #2
STEP 9   0.0 =   Basic Dimension of 125

The sum of these tolerances is 3.4. So, the maximum gap is:
```
  16.0
+  3.4
+19.4 = MAX GAP
```

The logic behind this loop route was to proceed from the left edge of the gap through all features that have an effect on the MAX GAP size, to the right edge of the gap. The route went left or right, as appropriate so as to involve all pertinent features, until the loop was complete. The pertinent features (in this view) were the four holes and the two pins passing through them.

Hole radii were used because the pertinent basic dimensions binding the gap to the holes, and the holes to each other, went to the hole centers. The full pin diameters were used because they were sandwiched between (jammed against the opposing sides) the holes. The basic dimensions were used to allow a route from the left side of the maximum gap to the center of Hole #1, then from the center of Hole #3 to the center of Hole #4, then from the center of Hole #3 to the right side of the maximum gap.

The route was:

- **Step 1** led from the left side of the MAX GAP to the center of Hole #1 (-125mm).
- **Step 2** led from the center of Hole #1 to the right side of Hole #1 (+3mm).
- **Step 3** led from the right side of the pin to the left side of the pin (-3mm).
- **Step 4** led from the left side of Hole #3 to the center of Hole #3 (+3mm).
- **Step 5** led from the center of Hole #3 to the center of Hole #4 (+260mm).
- **Step 6** led from the center of Hole #4 to the right side of Hole #4 (+3mm).
- **Step 7** led from the right side of the pin to the right side of the pin (-3mm).
- **Step 8** led from the left side of Hole #2 to the center of Hole #2 (+3mm).
- **Step 9** led from the center of Hole #2 to the right side of the gap (a right to left route, though) (-125mm).

If you examine the route taken, you can see that, with the parts shoved to create the MAX GAP, this is the only logical route to take. It gets you from the left side of the gap to the right side of the gap by taking the only known route of numbers.

CHAPTER 5
EXERCISES

Exercise 5-1

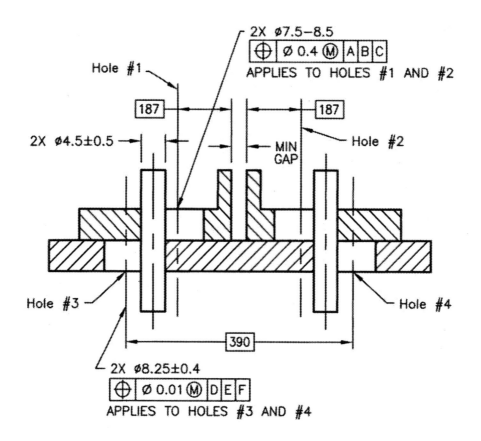

Exercise 5-1, Worksheet #1

Resultant Condition of Hole #1= _____
Virtual Condition of Hole #1= _____

Resultant Condition of Hole #2 _____
Virtual Condition of Hole #2 _____

Resultant Condition of Hole #3 _____
Virtual Condition of Hole #3 _____

Resultant Condition of Hole #4 _____
Virtual Condition of Hole #4 _____

  Res. Cond. Hole #1=   Res. Cond. Hole #1=
+ Virt. Cond. Hole #1= _____   - Virt. Cond. Hole #1= _____
      Sum   Difference

Sum and Difference Divided by 2= _____ =Mean Dia. $\pm$ Mean Tol.
Mean Dia. $\pm$ Mean Tol. Divided by 2= _____ =Mean Radius $\pm$ Mean Tol.

  Res. Cond. Hole #2=   Res. Cond. Hole #2=
+ Virt. Cond. Hole #2= _____   - Virt. Cond. Hole #2= _____
      Sum   Difference

Sum and Difference Divided by 2= _____ =Mean Dia. $\pm$ Mean Tol.
Mean Dia. $\pm$ Mean Tol. Divided by 2= _____ =Mean Radius $\pm$ Mean Tol.

  Res. Cond. Hole #3=   Res. Cond. Hole #3=
+ Virt. Cond. Hole #3= _____   - Virt. Cond. Hole #3= _____
      Sum   Difference

Sum and Difference Divided by 2= _____ =Mean Dia. $\pm$ Mean Tol.
Mean Dia. $\pm$ Mean Tol. Divided by 2= _____ =Mean Radius $\pm$ Mean Tol.

  Res. Cond. Hole #4=   Res. Cond. Hole #4=
+ Virt. Cond. Hole #4= _____   - Virt. Cond. Hole #4= _____
      Sum   Difference

Sum and Difference Divided by 2= _____ =Mean Dia. $\pm$ Mean Tol.
Mean Dia. $\pm$ Mean Tol. Divided by 2= _____ =Mean Radius $\pm$ Mean Tol.

Exercise 5-1, Worksheet #2

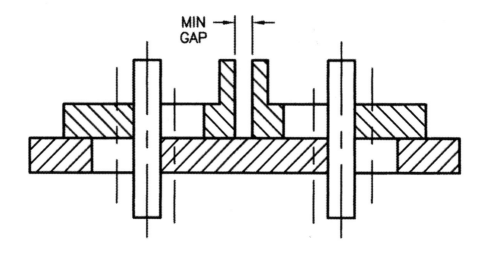

Loop Diagram

Right to Left	Left to Right	±Tol	
−	+		
			Totals
			=MIN GAP

63

Exercise 5-1, Worksheet #3

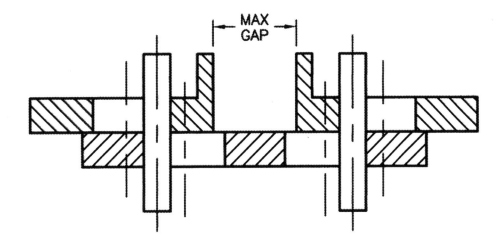

Loop Diagram

Right to Left	Left to Right	±Tol	
−	+		
			Totals
			=MAX GAP

Chapter 6

TOLERANCE STACK-UP ANALYSIS

•For a Fixed Fastener Assembly using Geometric Tolerances

•For Maximum Overall Dimension for a Crankshaft Assembly [Factors vs. Non-Factors]

- **Lesson Objectives:**
    In Chapter 6, for geometrically-toleranced fixed fastener assemblies, you will:
    - Calculate overall housing requirements--minimums and maximums.
    - Calculate MIN and MAX GAPS within the assembly.
    - Work with both a stationary assembly and a rotating assembly.
    - Calculate boundaries using a variety of geometric characteristics.

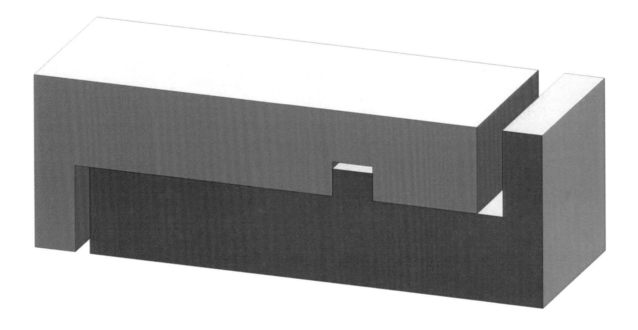

Chapter 6a
A Fixed Fastener Assembly using Geometric Tolerances

FIGURE 6-1

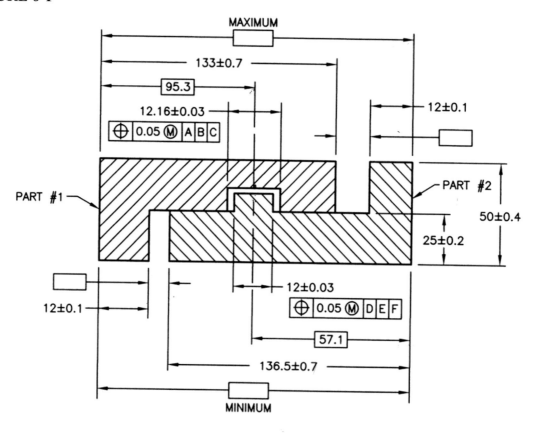

The analysis of this assembly will require calculating not only minimum and maximum gaps but also minimum and maximum dimensions of the assembly. To begin, we will calculate all pertinent inner and outer boundaries of geometrically toleranced features. Since the tab and slot have both been controlled at MMC, they will each have one constant boundary (virtual condition) and one non-constant boundary (resultant condition). The calculations are as follows:

FIGURE 6-2

Resultant Condition of Slot =

Virtual Condition of Slot =

Resultant Condition of Tab =

Virtual Condition of Tab =

Resultant Condition Slot − Virtual Condition Slot	= =	Virtual Condition Tab − Resultant Condition Tab	= =
Difference	=	Difference	=
1/2 Difference Slot	=	1/2 Difference Tab	=
Resultant Condition Slot + Virtual Condition Slot	= =	Virtual Condition Tab + Resultant Condition Tab	= =
Sum	=	Sum	=
1/2 Sum of R.C. and V.C. of Slot	=	1/2 Sum of V.C. and R.C. of Tab	=
1/2 Sum ±1/2 Diff. of Slot	=	1/2 Sum ±1/2 Diff. of Tab	=
1/2 of 1/2 Sum ±1/2 of 1/2 Diff. Slot	=	1/2 of 1/2 Sum ±1/2 of 1/2 Diff. Tab	=

Slot

　　12.13 = MMC Slot　　　　　　　　　　　　12.19 = LMC Slot
　　− 0.05 = Geo. Tol. at MMC　　and　　+ 0.11 = Geo. Tol. at LMC
　　12.08 = Virtual Condition Slot　　　　　12.30 = Resultant Condition Slot

Tab

　　12.03 = MMC Tab　　　　　　　　　　　　11.97 = LMC Tab
　　+ 0.05 = Geo. Tol. at MMC　　and　　− 0.11 = Geo. Tol. at LMC
　　12.08 = Virtual Condition Tab　　　　　11.86 = Resultant Condition Tab

Next, we find the plus and minus tolerance by finding the difference between the resultant and virtual conditions and dividing by 2. So,

　　Resultant Condition Slot　=　12.30
　　− Virtual Condition Slot　　=　12.08　　　then　　　$\frac{0.22}{2} = 0.11$
　　Difference　　　　　　　　=　 0.22

And,

　　Virtual Condition Tab　　　=　12.08
　　− Resultant Condition Tab　=　11.86　　　then　　　$\frac{0.22}{2} = 0.11$
　　Difference　　　　　　　　=　 0.22

Next we find the mean dimension to which we apply the plus and minus tolerance. So, we add the resultant and virtual conditions and divide by 2.

　　Resultant Condition Slot　=　12.30
　　+ Virtual Condition Slot　　=　12.08　　　then　　　$\frac{24.38}{2} = 12.19$
　　Sum　　　　　　　　　　　=　24.38

And,

Virtual Condition Tab	=	12.08
+ Resultant Condition Tab	=	11.86
Sum	=	23.94

then

$$\frac{23.94}{2} = 11.97$$

So, expressed as an equal bilateral toleranced dimension, the slot would be 12.19±0.11 and the tab is 11.97±0.11. Expressed as a radius, each would be:

$$\frac{12.19}{2} = 6.095 \quad \text{and} \quad \frac{0.11}{2} = 0.055$$

So; 6.095±0.055.

$$\frac{11.97}{2} = 5.985 \quad \text{and} \quad \frac{0.11}{2} = 0.055$$

So; 5.985±0.055.

FIGURE 6-3 [Virtual and Resultant Condition Calculations for Tolerance Stack-Up Analysis for Slot and Tab - Summary]

Resultant Condition of Slot = 12.19+0.11=12.3

Virtual Condition of Slot = 12.13−0.05=12.08

Resultant Condition of Tab = 11.97−0.11=11.86

Virtual Condition of Tab = 12.03+0.05=12.08

Resultant Condition Slot − Virtual Condition Slot	= 12.30 = 12.08	Virtual Condition Tab − Resultant Condition Tab	= 12.08 = 11.86
Difference	= 0.22	Difference	= 0.22
1/2 Difference Slot	= 0.11	1/2 Difference Tab	= 0.11
Resultant Condition Slot + Virtual Condition Slot	= 12.30 = 12.08	Virtual Condition Tab + Resultant Condition Tab	= 12.08 = 11.86
Sum	= 24.38	Sum	= 23.94
1/2 Sum of R.C. and V.C. of Slot	= 12.19	1/2 Sum of V.C. and R.C. of Tab	= 11.97
1/2 Sum ±1/2 Diff. of Slot	= 12.19±0.11	1/2 Sum ±1/2 Diff. of Tab	= 11.97±0.11
1/2 of 1/2 Sum ±1/2 of 1/2 Diff. Slot	= 6.095±0.055	1/2 of 1/2 Sum ±1/2 of 1/2 Diff. Tab	= 5.985±0.055

Now, we are ready to begin calculating the minimum overall dimension. To do this, we must visualize the parts being pushed together to create the smallest overall assembly dimension. This means the left side of the tab is pushed up against the left side of the slot. For this analysis, they are considered in the same location.

To run the loop analysis to calculate the minimum overall dimension, we must go: 1) from the left edge of Part #1 (the top part) to the center of the slot, 2) then back to the left edge of the slot and tab, 3) then back to the middle of the tab, and 4) then to the right edge of Part #2 (the bottom part).

FIGURE 6-4 [Calculate Minimum Overall Dimension]

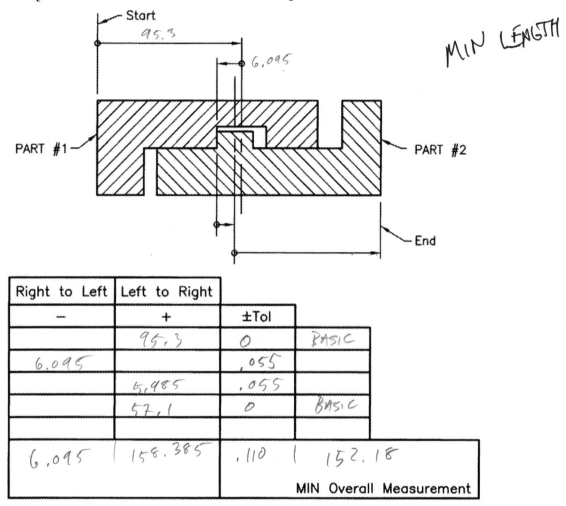

Right to Left	Left to Right	±Tol	
−	+		
	95.3	0	BASIC
6.095		.055	
	5.985	.055	
	57.1	0	BASIC
6.095	158.385	.110	152.18
			MIN Overall Measurement

STEP 1:
Start at the left edge of Part #1 and go right to the center of the slot (+93.300).
STEP 2:
Go left to the left edge of the slot (-6.095).
STEP 3:
Go right from the left edge of the slot and tab to the center of the tab (+5.985).
STEP 4:
Go right from the center of the tab to the right edge of Part #2 (+57.100).

Now we total the negatives, which consist of only one number, -6.096, and we total the positives:

```
   +95.300
+   +5.985
+  +57.100
  +158.385
```

The tolerances of all dimensions are added:

Tolerances

```
  0.000 = Basic Dimension of 95.300
  0.055 = Slot
  0.055 = Tab
+ 0.000 = Basic Dimension of 57.100
  0.110 = Total Tolerance
```

The negative 6.095 is added to the positive total of 158.385:

```
  +158.385
+   -6.095
  +152.290
```

The tolerance total is then subtracted from the 152.290 to get the MIN overall dimension.

```
  152.290
   -0.110
  152.180 = MIN Overall Measurement
```

FIGURE 6-5

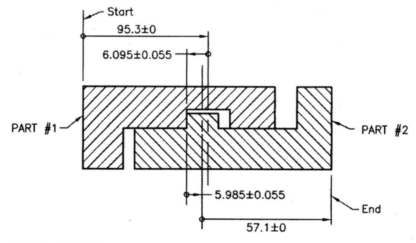

Right to Left	Left to Right	±Tol	
	95.300	0.000	Basic Dim
6.095		0.055	Slot
	5.985	0.055	Tab
	57.100	0.000	Basic Dim
6.095	158.385	0.110	Totals

```
  158.385        152.290
-   6.095      -   0.110
  152.290        152.18  MIN Overall Measurement
```

We then go to the maximum overall dimension. To do this loop analysis, we must visualize Part #1 and Part #2 being pulled out so as to create the maximum overall dimension for the two-part assembly. When this is done, the right side of the slot and tab will become one since they will be against one another.

FIGURE 6-6 [Calculate Maximum Overall Dimension]

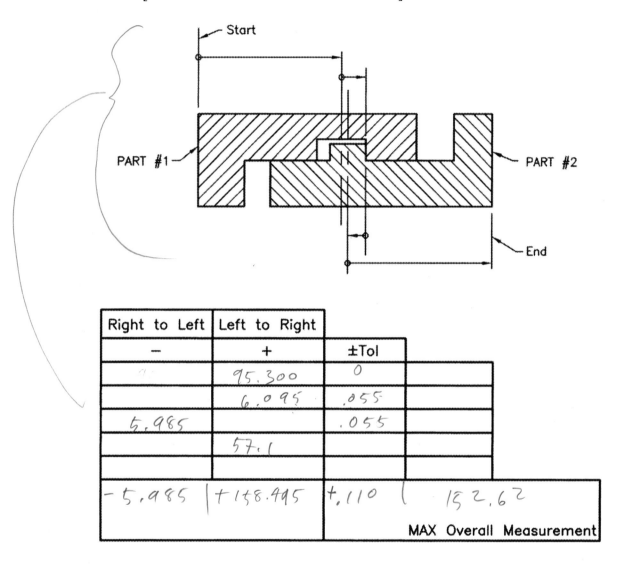

Right to Left	Left to Right	±Tol	
−	+		
	95.300	0	
	6.095	.055	
5.985		.055	
	57.1		
−5.985	+158.495	+.110	152.62
			MAX Overall Measurement

71

FIGURE 6-7 [Calculate Maximum Overall Dimension]

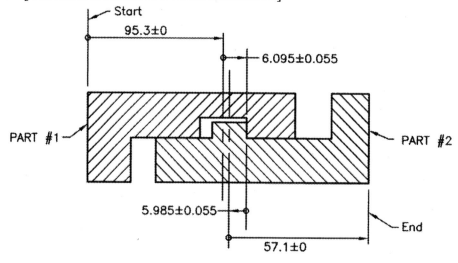

Right to Left	Left to Right	±Tol	
—	+		
	95.300	0.000	Basic Dim
	6.095	0.055	Slot
5.985		0.055	Tab
	57.100	0.000	Basic Dim
5.985	158.495	0.110	Totals

```
  158.495          152.51
-   5.985        + _0.11_
  152.510          152.62   MAX Overall Measurement
```

We now must start at the left edge of the part and proceed right (positive) the basic dimension of 95.300 to the center of the slot. Then go right again (positive) to the right edge of the slot 6.095. The loop then reverses and proceeds left (negative) to the center of the tab 5.985, then reverses again and goes right (positive) to the right edge of the assembly the basic dimension of 57.100.

STEP 1:
Start at the left edge of Part #1 and go right to the center of the slot (+95.300).
STEP 2:
Go right to the right edge of the slot (+6.095).
STEP 3:
Go left to the center of the tab (-5.985).
STEP 4:
Go right to the right edge of Part #2 (+57.100).

Now we total all the positives:
```
   +95.300
+   +6.095
+  +57.100
  +158.495
```

We then add the only negative number:
   +158.495
   + -5.985
   +152.510

The tolerances are then totaled:

Tolerances
   0.000 = Basic Dimension of 95.300
   0.055 = Slot
   0.055 = Tab
   0.000 = Basic Dimension of 57.100
   0.110 = Total Tolerance

The total tolerance is then added to the 152.510 to get the MAX overall dimension:
   152.51
   + 0.11
   152.62 = MAX overall dimension

FIGURE 6-8 [Calculate Minimum Gap Lower Left]

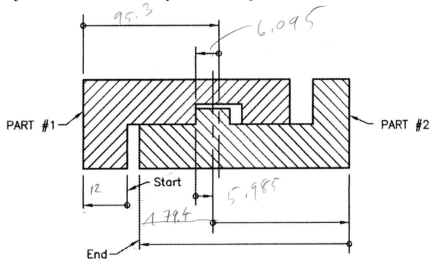

Right to Left	Left to Right	±Tol	
—	+		
12		0.1	Wall
	95.3	0	Basic Dim
6.095		.055	Slot
	5.985	.055	Tab
	57.1	0	Basic Dim
136.5		.70	Overall Dim
154.595	158.385	.91	Totals
	= 3.79	±.91 = 2.88	
		MIN GAP	

73

To calculate the minimum gap at the lower left of the assembly, we first visualize Part #1 and Part #2 pushed together to create the minimum gap. This will mean the left edge of the slot will contact the left of the tab. They share the same location in the assembly.

FIGURE 6-9 [Calculate Minimum Gap Lower Left]

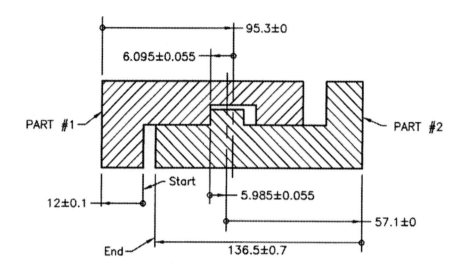

Right to Left	Left to Right	±Tol	
−	+		
12.000		0.100	Wall
	95.300	0.000	Basic Dim
6.095		0.055	Slot
	5.985	0.055	Tab
	57.100	0.000	Basic Dim
136.500		0.700	Overall Dim
154.595	158.385	0.910	Totals
158.385 −154.595 3.790		3.79 −0.91 2.88	MIN GAP

Since we are calculating the minimum gap at the lower left, we will begin the loop at the left side of the gap (on Part #1) and go left (negative) 12mm to the edge of Part #1. Then we will go right (positive) to the center of the slot 95.300. Then we go left (negative) to the edge of the slot 6.095. Now we go right (positive) to the center of the tab 5.985. The loop continues right (positive) to the right edge of Part #2 a distance of 57.100. The loop reverses and goes left (negative) for the last step to the edge of the gap being calculated a distance of 136.500.

STEP 1:
Go left from the left edge of the minimum gap to the left edge of Part #1 (-12.000).
STEP 2:
Go right from the edge of Part #1 to the center of the slot (+95.300).
STEP 3:
Go left to the left edge of the slot (-6.095).
STEP 4:
Go right to the center of the tab (+5.985).
STEP 5:
Go right to the right edge of Part #2 (+57.100).
STEP 6:
Go left to the left edge of Part #2 (-136.500).

The negatives are added:
$$\begin{aligned} &-12.000 \\ + \ &-6.095 \\ + \ &\underline{-136.500} \\ &-154.595 \end{aligned}$$

The positives are added:
$$\begin{aligned} &+95.300 \\ + \ &+5.985 \\ + \ &\underline{+57.100} \\ &+158.385 \end{aligned}$$

The negative and positive totals are added:
$$\begin{aligned} &+158.385 \\ + \ &\underline{-154.595} \\ &+3.790 \end{aligned}$$

The tolerances are totaled:

Tolerances

    0.100 = Wall
    0.000 = Basic Dimension of 95.300
    0.055 = Slot
    0.055 = Tab
    0.000 = Basic Dimension of 57.100
  + 0.700 = Overall Dimension of Part #2
    0.910 = Total ± Tolerance

The total tolerance is subtracted from the 3.79 to get the MIN Gap:
    3.79
  - 0.91
    2.88 = MIN Gap (at lower left)

FIGURE 6-10

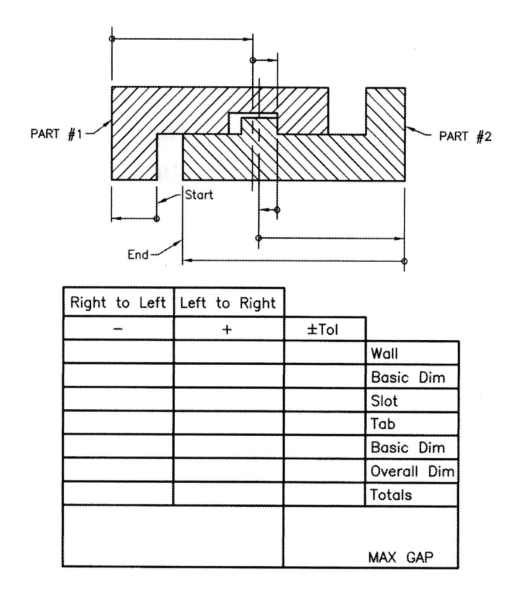

To **calculate the maximum gap at the lower left**, we must visualize the parts pushed in a fashion that opens the gap to its widest. Part #2 can be pushed to the right until the right side of the tab is against the right side of the slot. The gap is as wide as it can be.

The loop will begin at the left side of the gap, and then proceed left to the left edge of Part #1, then right to the center of the slot. Go right again to the right side of the slot, then left to the center of the tab, then right to the right edge of Part #2, then left all the way back to the edge of Part #2 and the end of the gap.

FIGURE 6-11

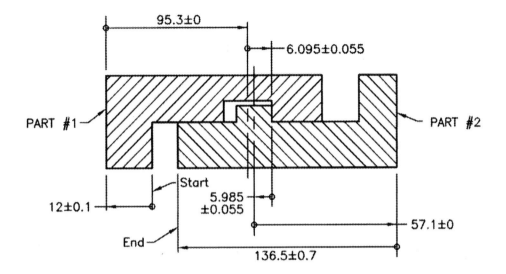

Right to Left	Left to Right	±Tol	
−	+		
12.000		0.100	Wall
	95.300	0.000	Basic Dim
	6.095	0.055	Slot
5.985		0.055	Tab
	57.100	0.000	Basic Dim
136.500		0.700	Overall Dim
154.485	158.495	0.910	Totals

```
 158.495        4.01
-154.485       +0.91
   4.010        4.92    MAX GAP
```

STEP 1:
From the left side of the gap, go left to the left edge of Part #1 (-12.000).
STEP 2:
Go right to the center of the slot (+95.300).
STEP 3:
Go right to the right side of the slot (+6.095).
STEP 4:
Go left to the center of the tab (-5.985).
STEP 5:
Go right to the right edge of Part #2 (+57.100).
STEP 6:
Go left to the left edge of Part #2 and the end of the gap (-136.500).

We total all the negatives:
```
     -12.000
  +   -5.985
  + -136.500
    -154.485
```

Now we total all the positives:
```
     +95.300
  +   +6.095
  +  +57.100
    +158.495
```

We then add the negative total to the positive total:
```
    +158.495
  + -154.485
      +4.010
```

We total all the tolerances:

Tolerances
```
    0.100 = Wall
    0.000 = Basic Dimension of 95.300
    0.055 = Slot
    0.055 = Tab
    0.000 = Basic Dimension of 57.100
  + 0.700 = Overall Dimension of 136.500
    0.910 = Total Tolerance
```

To calculate the maximum gap, we add the total tolerance to the 4.01:
```
    4.01
  + 0.91
    4.92 = MAX GAP
```

FIGURE 6-12 [Calculate Maximum Gap Upper Right]

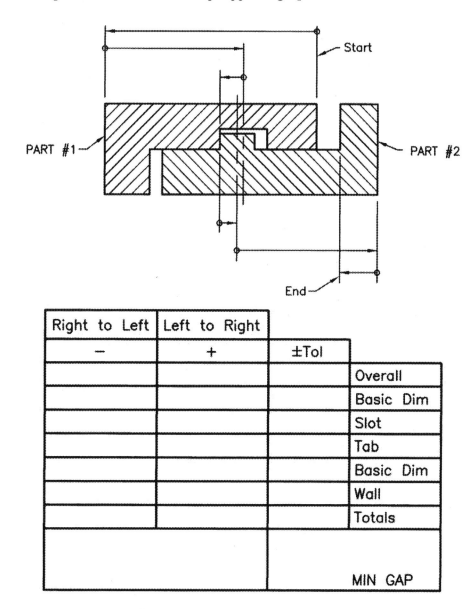

To calculate the minimum gap at the upper right corner, we must visualize the parts pushed together until the left edge of the tab is against the left edge of the slot.

To begin, we start at the left edge of the gap on Part #1 and proceed left to the left edge of the part, then we go right to the center of the slot, then left to the left edge of the slot, then right to the center of the tab, then again to the right edge of Part #2, then left to the right side of the gap (the left side of the wall).

FIGURE 6-13 [Explanation for FIGURE 6-12]

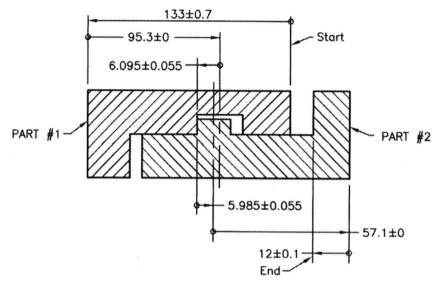

Right to Left	Left to Right	±Tol	
−	+		
133.000		0.700	Overall
	95.300	0.000	Basic Dim
6.095		0.055	Slot
	5.985	0.055	Tab
	57.100	0.000	Basic Dim
12.000		0.100	Wall
151.095	158.385	0.910	Totals

```
 158.385       7.29
-151.095      -0.91
   7.290       6.38   MIN GAP
```

STEP 1:
From the left edge of the gap, go left to the left edge of Part #1 (-133.000).
STEP 2:
Go right to the center of the slot (+95.300).
STEP 3:
Go left to the left edge of the slot (-6.095).
STEP 4:
Go right to the center of the tab (+5.985).
STEP 5:
Go right again to the right edge of Part #2 (+57.100).
STEP 6:
Go left to the end of the gap (the right edge of the wall) (-12.000).

We total all the negatives:
    -133.000
+   -6.095
+ -12.000
   -151.095

Now we total all the positives:
   +95.300
+ +5.985
+ +57.100
 +158.385

We then add the positive and negative totals:
  +158.385
+ -151.095
    +7.290

We total all the tolerances:
Tolerances
0.700 = Overall (Tolerance on 133.000)
0.000 = Basic Dimension of 95.300
0.055 = Slot
0.055 = Tab
0.000 = Basic Dimension of 57.100
+ 0.100 = Wall (Tolerance on 12.000)
0.910 = Total Tolerance

To calculate the minimum gap, we subtract the total tolerance from the 7.29:
  7.29
- 0.91
  6.38 = MIN GAP

FIGURE 6-14 [Calculate Maximum Gap Upper Right]

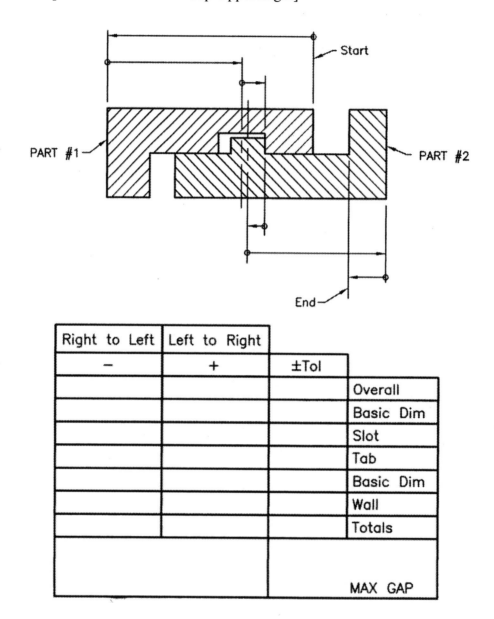

Right to Left	Left to Right	±Tol	
−	+		
			Overall
			Basic Dim
			Slot
			Tab
			Basic Dim
			Wall
			Totals
			MAX GAP

To calculate the maximum gap in the upper right corner, we must visualize the parts pulled apart to create the widest gap. With this done, the right side of the tab will rest against the right side of the slot. To begin, we start at the left side of the gap (the right side of Part #1) and go left to the left side of Part #1, then we go right to the center of the slot, then right again to the right side of the slot, then left to the center of the tab, then right to the right edge of Part #2, then left to the end of the loop (the right side of the gap--the left side of the wall).

FIGURE 6-15 [Explanation for FIGURE 6-14]

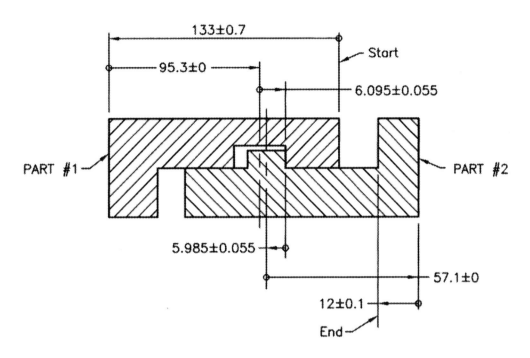

Right to Left	Left to Right	±Tol	
−	+		
133.000		0.700	Overall
	95.300	0.000	Basic Dim
	6.095	0.055	Slot
5.985		0.055	Tab
	57.100	0.000	Basic Dim
12.000		0.100	Wall
150.985	158.495	0.910	Totals

```
 158.495      7.51
-150.985     +0.91
   7.510      8.42   MAX GAP
```

STEP 1:
Begin at the right edge of Part #1 and go left to the left edge of Part #1 (-133.000).
STEP 2:
Go right to the center of the slot (+95.300).
STEP 3:
Continue right to the right edge of the slot (+6.095).
STEP 4:
Go left to the center of the tab (-5.985).
STEP 5:
Go right to the right edge of Part #2 (+57.100).

STEP 6:
Go left to the end of the loop (right side of the gap--left edge of the wall) (-12.000).

We total all the negatives:
```
   -133.000
+   -5.985
+  -12.000
  -150.985
```

Now we total all the positives:
```
   +95.300
+   +6.095
+  +57.100
  +158.495
```

We then add the positive and negative totals:
```
  +158.495
+ -150.985
    +7.510
```

We total all the tolerances:

Tolerances
```
  0.700 = Overall (Tolerance on 133.000)
  0.000 = Basic Dimension of 95.300
  0.055 = Slot
  0.055 = Tab
  0.000 = Basic Dimension of 57.100
+ 0.100 = Wall (Tolerance on 12.000)
  0.910 = Total Tolerance
```

To calculate the maximum gap, we add the total tolerance to the 7.51:
```
   7.51
+  0.91
   8.42 = MAX GAP
```

Chapter 6b
Maximum Overall Dimension For Crankshaft Assembly
[Factors vs. Non-Factors]

FIGURE 6-16 [Assembly]

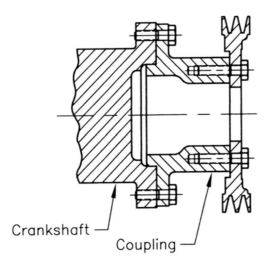

FIGURE 6-17 Detail Drawings

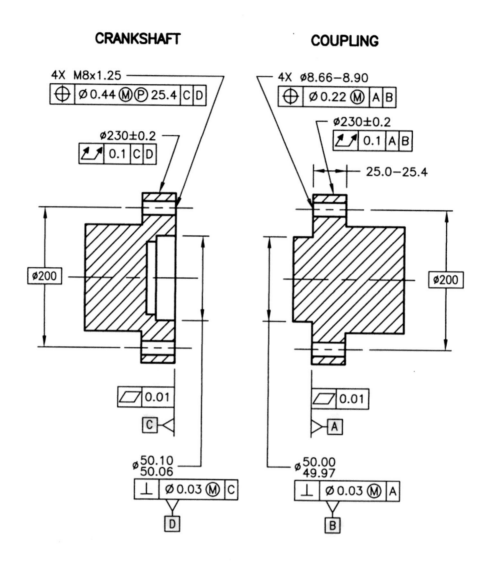

Depending on what one is trying to calculate, it may be determined that some of the features on the parts in the assembly do not factor into the equation. For example, if we are to determine the maximum overall assembly size for the diameters, the threaded and clearance holes would not be factors. In their tightest fits, these features would be line fits of Ø8.44 (8.66 – 0.22 for the clearance holes and 8.00 + 0.44 for the screws mounted into the threaded holes).

The datum features (B and D) would also be line fits of Ø50.03 (50.06 - 0.03 for D and 50.00 + 0.03 for B). But in the LMC fits, the datum features have less clearance (Ø50.10 - 49.97 = Ø0.13) than the threaded features and their clearance holes (Ø8.90 - 8.00 = Ø0.90).

This means (having less potential clearance) the datum features determine (are the factors for) the maximum the outside diameters may be offset from one another. Therefore, in this type of calculation, the threaded holes (and screws) and the clearance holes will not be considered.

FIGURE 6-18

Step 1:

$$230.2 = \text{MMC}$$
$$+\ \ 0.1 = \text{Geometric Tolerance at MMC}$$
$$\overline{\varnothing 230.3 = \text{Outer Boundary of Shaft}}$$

$$\varnothing 230.3/2 = \text{Radius of } 115.15$$

Step 2:

Coupling Datum Feature B LMC=Ø49.97

Ø49.97/2=Radius of 24.985

Step 3:

CrankshaftDatum Feature D LMC=Ø50.10

Ø50.10/2=Radius of 25.05

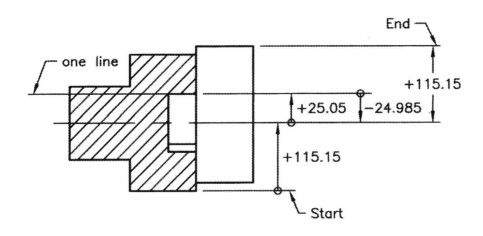

−	+
	115.150
	25.050
24.985	
	115.150
24.985	255.35	Totals

```
  255.350
-  24.985
  -------
  230.365 = MAX Overall Dimension
```

[NOTE: Since this is a rotating assembly, the housing required to contain just these two parts **while rotating** would be a minimum of 115.15 x 2 plus (25.050 minus 24.985) x 2. This would be 230.300+0.130 = **230.43**.

87

FIGURE 6-19 [Another Way of Looking at It]

Step 1:
        230.2=MMC
  +   0.1=Geometric Tolerance at MMC
     Ø230.3=Outer Boundary of Shaft

     Ø230.3/2=Radius of 115.15

Step 2:
     Coupling Datum Feature B LMC=Ø49.97

     Ø49.97/2=Radius of 24.985

Step 3:
     CrankshaftDatum Feature D LMC=Ø50.10

     Ø50.10/2=Radius of 25.05

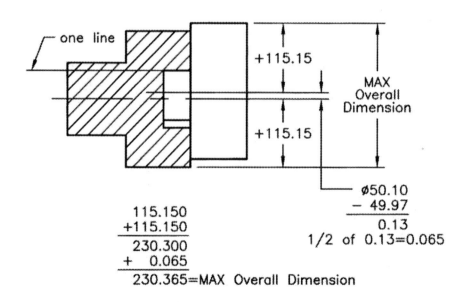

115.150
+115.150
─────
230.300
+ 0.065
─────
230.365=MAX Overall Dimension

Ø50.10
− 49.97
─────
0.13
1/2 of 0.13=0.065

NOTE: Since a MIN Overall Dimension would have the crankshaft and coupling perfectly centered, the MIN can be calculated by just adding the 2 radii of the parts' LMC dimensions. So, the MIN is 114.9 plus 114.9 equals 229.8. If the parts had different LMC diameter dimensions, the largest would be used as the MIN Overall Dimension of the two parts once assembled.

Again, although 230.365 is the MAX overall dimension, the minimum housing requirement **while rotating** would be 0.065 larger. 230.365 + 0.065 = **230.43**.

CHAPTER 6
EXERCISES

Exercise 6-1

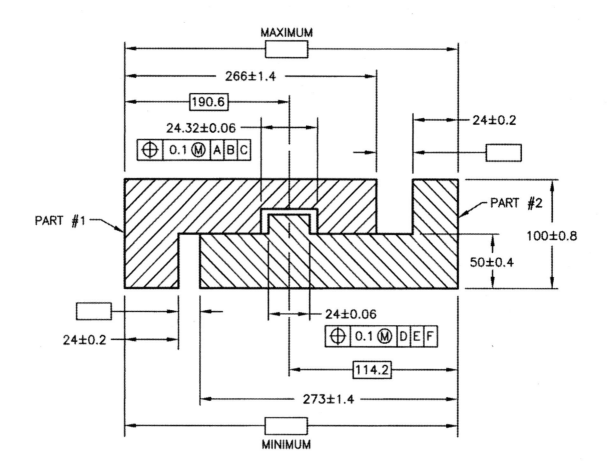

Exercise 6-1, Worksheet #1

Resultant Condition of Slot =

Virtual Condition of Slot =

Resultant Condition of Tab =

Virtual Condition of Tab =

Resultant Condition Slot = − Virtual Condition Slot =	Virtual Condition Tab = − Resultant Condition Tab =
Difference =	Difference =
1/2 Difference Slot =	1/2 Difference Tab =
Resultant Condition Slot = + Virtual Condition Slot =	Virtual Condition Tab = + Resultant Condition Tab =
Sum =	Sum =
1/2 Sum of R.C. and V.C. of Slot =	1/2 Sum of V.C. and R.C. of Tab =
1/2 Sum ±1/2 Diff. of Slot =	1/2 Sum ±1/2 Diff. of Tab =
1/2 of 1/2 Sum ±1/2 of 1/2 Diff. Slot =	1/2 of 1/2 Sum ±1/2 of 1/2 Diff. Tab =

Exercise 6-1, Worksheet #2

PROBLEM: Calculate the minimum overall dimension.

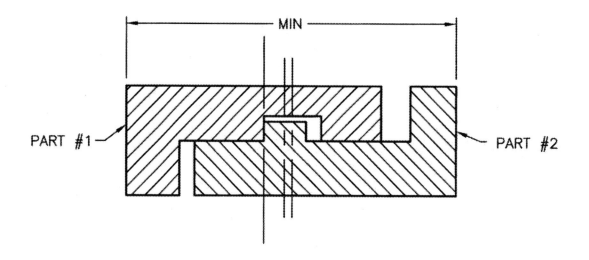

Right to Left	Left to Right	±Tol	
−	+		
		Totals	
		=MIN Overall Measurement	

92

Exercise 6-1, Worksheet #3

PROBLEM: Calculate the maximum overall dimension.

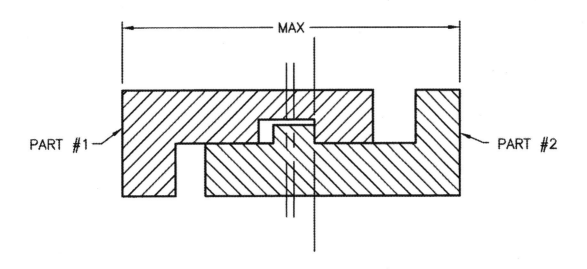

Right to Left	Left to Right	±Tol	
−	+		
			Totals
			=MAX Overall Measurement

Exercise 6-1, Worksheet #4

PROBLEM: Calculate the minimum gap lower left.

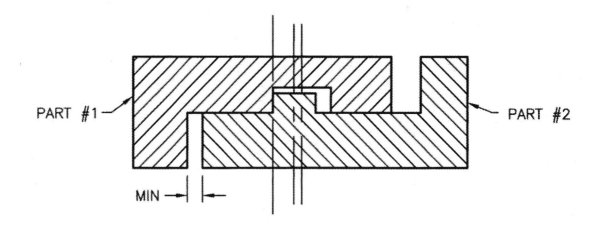

Right to Left	Left to Right	±Tol	
−	+		
			Totals
			=MIN GAP

Exercise 6-1, Worksheet #5

PROBLEM: Calculate the maximum gap lower left.

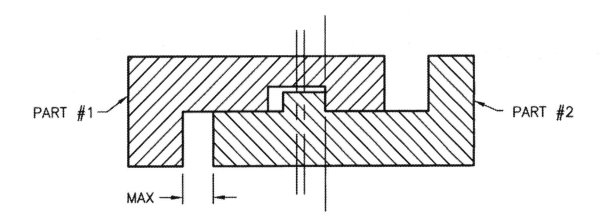

Right to Left	Left to Right	±Tol	
−	+		
			Totals
			=MAX GAP

95

Exercise 6-1, Worksheet #6

PROBLEM: Calculate the minimum gap upper right.

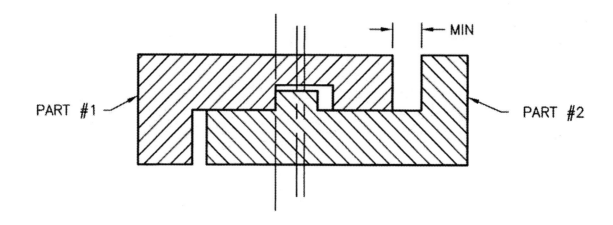

Right to Left	Left to Right	±Tol	
−	+		
			Totals
			=MIN GAP

Exercise 6-1, Worksheet #7

PROBLEM: Calculate the maximum gap upper right.

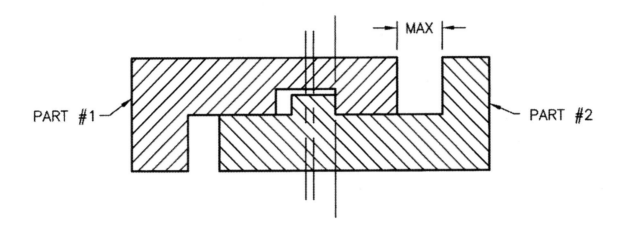

Right to Left	Left to Right	±Tol	
−	+		
			Totals
			=MAX GAP

Exercise 6-2 Worksheet #1

PROBLEM: Determine the maximum overall assembly dimension required to contain both the crankshaft and the coupling diameters.

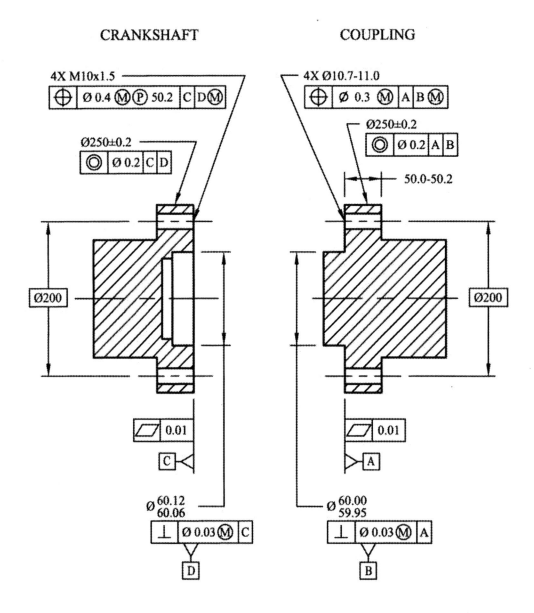

Chapter 7

TOLERANCE STACK-UP ANALYSIS FOR A RAIL ASSEMBLY FIXED FASTENER SITUATION

•Lesson Objective:
In Chapter 7, you will:
- Calculate pertinent boundaries for threaded features.
- Work with multiple geometric controls on a single feature.
- Determine the effects of projected tolerance zones on a stack-up.
- Determine which geometric tolerances affect the analysis you are performing.
- Calculate clearance and interference.
- Use logic in your calculations.
- Use product knowledge, experience and assembly conditions in your analysis.

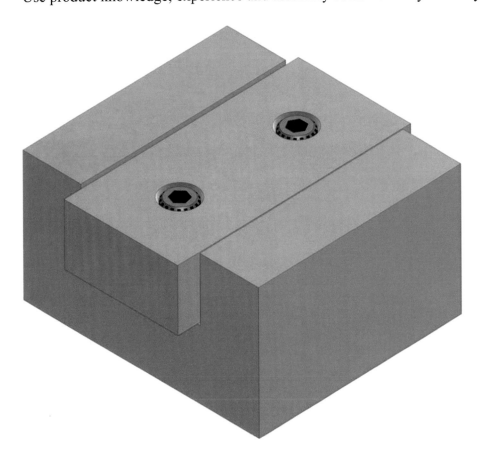

Chapter 7
Rail Assembly Fixed Fastener

FIGURE 7-1 (Rail Assembly)

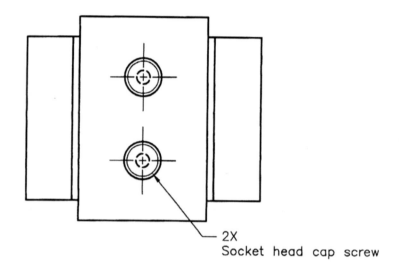

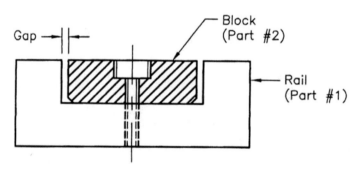

FIGURE 7-2 [Rail]

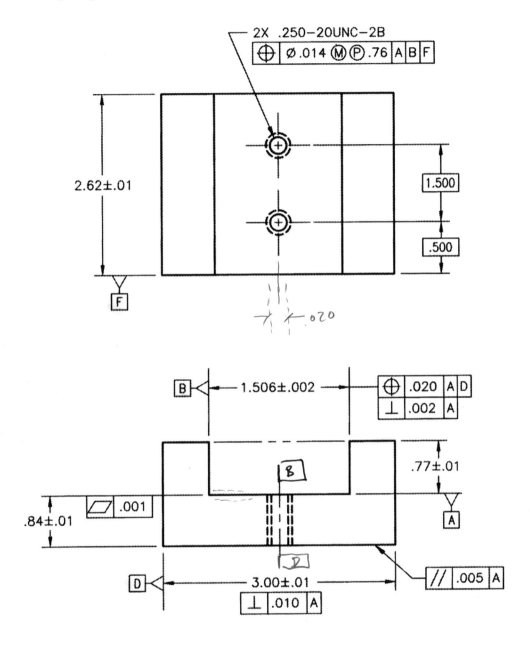

Note: LMC of screw is Ø.2408; MMC is Ø.250.

Since datum feature B is referenced in the feature control frame (position tolerance) on the threaded holes, the position of datum feature B is not a factor and is excluded in this analysis. No matter where datum feature B is located, the threaded holes are centered to it. However, the perpendicularity tolerance on datum feature B does affect how the parts assemble, so it is included as a factor in this analysis.

FIGURE 7-3 [Block]

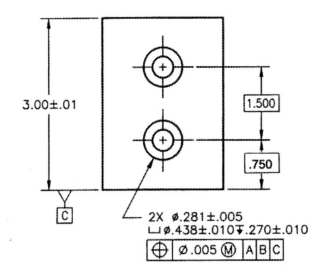

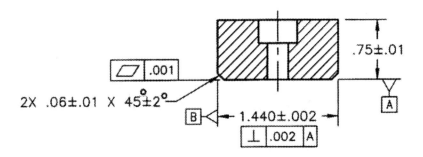

Remember: If your assumptions are wrong, your answer is wrong.

IMPORTANT RULES:

Correct routes give the largest MIN GAP and the smallest MAX GAP.

Run multiple routes to determine (for example) MIN GAP. The route that gives you the largest MIN GAP is the one that is correct, because it is the one that aligns (and stops the gap from being smaller) or hits first (and stops the gap from being smaller).

MAX GAP: The route that gives you the smallest MAX GAP is the one that is correct, because it is the one that hits first.

These routes assume your "one line" is the correct place where one part contacts another.

Always take the shortest route. It is easy to get a wrong answer if your route is not the shortest because it will accumulate error (tolerances) that should not be included in your calculation.

A "red flag" should be visualized when the MIN GAP calculated is a negative number. The MIN GAP may be a negative number sometimes, but at other times it is a physical impossibility. When worst mating boundaries of individual mating features show no interference, then shoving one part until it hits another reaps a MIN GAP of zero. So, if it is the clearance being used between mating features to create interference, when the inner boundary for holes and slots and outer boundaries for shafts and tabs show no interference, the parts don't have to interfere. The clearance can be used to optimize the assembly and allow the parts to mate, instead of being used to push one part into interference with another.

However, when a MIN GAP calculation reaps a negative number, the MIN GAP actually becomes the MAX INTERFERENCE. Still, remember not to accept this MAX INTERFERENCE as correct until you examine whether or not it is both physically possible and logical from an optimal assembly standpoint.

It should be remembered that even though individual "worst mating conditions" may show no interference, pushing parts-using clearance to make them hit one another, which is done in this approach, can make interference show up in more than one place. This just means that when multiple routes are run to determine the correct route, interference can be seen to occur in more than one place. Even though this is physically impossible if individual "worst mating conditions" are shown to be compatible, we know that the parts can often be shoved until one hits another somewhere.

Multiple MIN GAP calculations can be run for different places where interference is feared. When this is done, more than one MIN GAP can show interference instead of clearance. This "interference" is often misleading in multiple ways. The first way is that interference can't physically occur. This is true when one part hits another and the minimum gap is actually zero, not interference at all. The second way is that the GAP shown to have the most interference (the largest negative number) is the one that hits first when the parts are pushed. This means the smallest interference is not the most likely alignment feature. So, as will be shown in this unit, the mating features with the minimum clearance are the most likely to touch and create the "one line" (where one part contacts another) and the other set of features are most likely to have a GAP.

The minimum clearance is shown between the virtual conditions of the clearance holes (a diameter of .271) and the screws once mounted into their threaded holes (a diameter of .264). This minimum clearance is a diameter of .007. The difference between their least material conditions (.286 minus .2408) equals a diameter of .0452 (their maximum clearance).

The minimum clearance between the inner boundary of the cavity (1.502) and the outer boundary of the block (1.444) equals .058. The difference between their least material conditions (1.508 minus 1.438) equals .070 (their maximum clearance). The .070 is a number that should be remembered when calculating the MAX GAP. Other mating features can only subtract from that .070 MAX GAP. They can't add to it. This is often forgotten when the shortest route is not taken and inappropriate tolerances are factored in. This will be shown in later calculations in this unit.

Since they have the smallest clearances, the likely "one-line" (where one part contacts the other) for MIN GAP calculations is between the clearance holes and the screws once they are mounted into their threaded holes. If we were to calculate the MIN GAP between the clearance holes and the screws (once they are mounted into the threaded holes) using the same numbers that we used for the MIN GAP between the block and the rail, the interference would be shown to be much greater than the interference between the block and the rail (-.0335 vs. -.0081).

The rail assembly shown in FIGURES 7-1, 7-2 and 7-3 is a classic example of a fixed fastener assembly, with clearance holes in the block and threaded holes in the rail. When the screw passes through the clearance holes and is screwed into the threaded holes in the rail, the threaded holes fix the screw (grab it). So, together, these parts constitute a fixed fastener assembly condition. Even datum features B on both parts are considered a fixed fastener mating condition because the male feature B that is the block is given some of the geometric tolerance.

Whenever a male feature is given a geometric tolerance, in order for the worst mating conditions on the male and female to be compatible, the geometric tolerance must be divided between the mating features (parts). This assures us the condition is a fixed fastener assembly. If the full amount of the available geometric tolerance was used on both parts (without division), it would be categorized as a floating fastener condition. To do this without interference occurring, both parts would have to have clearance holes, and screws that pass through them and (together with nuts) bind the parts to one another.

FIGURE 7-4 [Virtual and Resultant Condition Calculations for Tolerance Stack-Up Analysis for Rail and Block]

Outer Boundary of Slot in Rail = _____

Inner Boundary of Slot in Rail = _____

Inner Boundary of Width of Block = _____

Outer Boundary of Width of Block = _____

Outer Boundary Slot = − Inner Boundary Slot = ──────────── Difference =	Outer Boundary Block = − Inner Boundary Block = ──────────── Difference =
1/2 Difference of Slot = _____	1/2 Difference of Block = _____
Outer Boundary Slot = + Inner Boundary Slot = ──────────── Sum =	Outer Boundary Block = + Inner Boundary Block = ──────────── Sum =
1/2 Sum of O.B. and I.B. of Slot = _____	1/2 Sum of O.B. and I.B. of Block = _____
1/2 Sum ± 1/2 Diff. of Slot = _____	1/2 Sum ± 1/2 Diff. of Block = _____
1/2 of 1/2 Sum ± 1/2 of 1/2 Diff. Slot = _____	1/2 of 1/2 Sum ± 1/2 of 1/2 Diff. Block = _____

TO BEGIN, we will determine the pertinent boundaries for the features that influence what we are trying to calculate. We will calculate MIN and MAX GAPS for the block and rail when assembled. So, the pertinent boundaries are for the slot in the rail, the width of the block, the screws--when mounted--in the threaded holes, and the clearance holes the screw must pass through. This calculation assumes that when the block is pushed toward the wall in the rail's slot, the screw hits the side of the clearance hole before the block hits the side of the slot in the rail. If this assumption is wrong, we could get a negative number for the minimum gap between the left edge of the block and the edge of the slot rail. If we calculate the MIN GAP and get a negative number, it might mean that the block won't fit into the cavity, or that the left edge of the block hits the edge of the rail before the screws that are mounted in the threaded holes hit the side of the clearance holes. This discovery would mean that all of our calculations were based on a false assumption, therefore our answer is wrong and, when pushed toward one another, there is no airspace between the edge of the block and the slot in the rail. That would mean that, using those numbers and that route, the GAP would actually be between the screws mounted in the threaded holes and the clearance holes in the block.

RAIL
So, let's begin the first route to calculate the MIN GAP shown (between the block and the slot in the rail) and determine if our assumption is correct. The slot in the rail has two geometric tolerances, but only the perpendicularity tolerance will effect how the block fits into the slot. This is because the threaded holes are positioned to the slot centerplane (datum B) no matter how out-of-position the slot is produced. Therefore, the positional tolerance of the slot is not a factor and will be ignored for this analysis. Since the geometric tolerance in the perpendicularity control is Regardless of Feature Size, to get the outer boundary of the slot, we add .002 to the LMC (1.508); to get the inner boundary, we subtract .002 from the MMC (1.504). The outer boundary of the slot is 1.510 and the inner boundary is 1.502. We subtract these boundaries to get the difference:

    1.510
  - 1.502
     .008   .008 ÷ 2 = .004. This is the plus and minus tolerance.

We then add 1.510 and 1.502 to get the sum. The sum of 3.012 is divided by 2 to find the mean dimension.

    1.510
  + 1.502
    3.012   3.012 ÷ 2 = 1.506, to which the plus and minus tolerance is given. 1.506±.004.

Displayed as a radius, both are divided by 2 again and we get:
.753 (1/2 of 1.506) ±.002 (1/2 of ±.004).

The same procedure is performed for the width of the block. 1.438 (LMC) minus the .002 perpendicularity tolerance gets us an inner boundary of 1.436 and 1.442 (MMC) plus the .002 perpendicularity tolerance gets us an outer boundary of 1.444. These are subtracted:

      1.444 = Outer Boundary Block
   - 1.436 = Inner Boundary Block
       .008 = Difference

The .008 difference is divided by 2 to give us .004. The boundaries are added and the sum divided by 2 to give us the mean dimension.

$$\begin{aligned} 1.444 &= \text{Outer Boundary Block} \\ + \underline{1.436} &= \underline{\text{Inner Boundary Block}} \\ 2.880 &= \text{Sum} \\ \frac{2.880}{2} &= 1.440 \text{ Mean Dimension} \end{aligned}$$

The tolerance is given, so it becomes a width that is 1.440±.004. To convert this width to a radius, both the dimension and the tolerance are divided by 2, which give us .720±.002.

FIGURE 7-5 [Virtual and Resultant Condition Calculations for Tolerance Stack-Up Analysis for Rail and Block]

Outer Boundary of Slot in Rail = 1.508 + .002 = 1.510

Inner Boundary of Slot in Rail = 1.504 − .002 = 1.502

Inner Boundary of Width of Block = 1.438 − .002 = 1.436

Outer Boundary of Width of Block = 1.442 + .002 = 1.444

SLOT	BLOCK
Outer Boundary Slot = 1.510 − Inner Boundary Slot = 1.502 Difference = .008	Outer Boundary Block = 1.444 − Inner Boundary Block = 1.436 Difference = .008
1/2 Difference of Slot = .004	1/2 Difference of Block = .004
Outer Boundary Slot = 1.510 + Inner Boundary Slot = 1.502 Sum = 3.012	Outer Boundary Block = 1.444 + Inner Boundary Block = 1.436 Sum = 2.880
1/2 Sum of O.B. and I.B. of Slot = 1.506	1/2 Sum of O.B. and I.B. of Block = 1.440
1/2 Sum ±1/2 Diff. of Slot = 1.506±.004	1/2 Sum ±1/2 Diff. of Block = 1.440±.004
1/2 of 1/2 Sum ±1/2 of 1/2 Diff. Slot = .753±.002	1/2 of 1/2 Sum ±1/2 of 1/2 Diff. Block = .720±.002

FIGURE 7-6

Inner Boundary of Screw Mounted in Rail =

Outer Boundary of Screw Mounted in Rail =

Outer Boundary of Hole in Block =

Inner Boundary of Hole in Block =

Outer Boundary Mounted Screw = − Inner Boundary Mounted Screw =	Outer Boundary Hole = − Inner Boundary Hole =
Difference =	Difference =
1/2 Difference Mounted Screw =	1/2 Difference Hole =
Outer Boundary Mounted Screw = + Inner Boundary Mounted Screw =	Outer Boundary Hole = + Inner Boundary Hole =
Sum =	Sum =
1/2 Sum of O.B. and I.B. Screw =	1/2 Sum of O.B. and I.B. Hole =
1/2 Sum ±1/2 Diff. Screw =	1/2 Sum ±1/2 Diff. Hole =
1/2 of 1/2 Sum ±1/2 of 1/2 Diff. Screw =	1/2 of 1/2 Sum ±1/2 of 1/2 Diff. Hole =

SCREWS

- Now the boundaries for the screws mounted in the screw holes are calculated. The screws take on the positional tolerance of the threaded holes, especially since the threaded holes project their tolerance zones into the screws. So, the screws and the screw holes will be treated as one, just as though they were shafts. The inner boundary of the screw mounted in the rail is calculated by subtracting the threaded holes' positional tolerance (Ø.014) from the LMC of the screw (Ø.2408, according to the Machinery's Handbook). So, Ø.2408
- .0140
Ø.2268

The outer boundary of the screw mounted in its threaded hole is the MMC (Ø.250) plus its positional tolerance (Ø.014) which equals Ø.264. The boundaries are subtracted to obtain the tolerance (difference):

.2640 = Outer Boundary Mounted Screw
- .2268 = Inner Boundary Mounted Screw
.0372 = Difference (tolerance)

Converted to a plus and minus tolerance,
$$\frac{.0372}{2} = .0186$$

The boundaries are then added and the sum divided by 2 to get the nominal dimension:

.2640 = Outer Boundary Mounted Screw
+ .2268 = Inner Boundary Mounted Screw
.4908 = Sum

$$\frac{.4908}{2} = .2454 = \text{Mean/Nominal Dimension}$$

The tolerance is applied: Ø.2454±.0186. This is then converted to a radius by dividing both the dimension and the tolerance by 2:

$$\frac{Ø.2454}{2} = R.1227 \quad \text{and} \quad \frac{Ø.0186}{2} = R.0093$$

So, the final radial dimension and tolerance to be used in the calculation is R.1227±.0093.

BLOCK

Now, the holes in the block need their boundaries calculated. The outer boundary is calculated by adding the LMC (Ø.286) and the geometric tolerance applicable at LMC (Ø.015) to equal Ø.301. The inner boundary is calculated by subtracting the applicable geometric tolerance at MMC (Ø.005) from the hole's MMC (Ø.276) to equal Ø.271.

The boundaries are subtracted to get the tolerance (difference):

.301 = Outer Boundary Holes
- .271 = Inner Boundary Holes
Ø.030 = Tolerance (difference)

And the tolerance (Ø.030) is divided by 2 to convert it to a plus and minus tolerance:
$$\frac{.030}{2} = .015$$

The boundaries are then added and divided by 2 to get a mean or nominal diameter:

Ø.301 = Outer Boundary Holes
+ Ø.271 = Inner Boundary Holes
Ø.572 = Sum

$$\frac{Ø.572}{2} = Ø.286 \text{ Nominal Dimension}$$

The tolerance is applied to the mean diameter and gets us Ø.286±.015. Then, both the dimension and the tolerance are divided by 2 to convert the diameters to radii. So,

$$\frac{Ø.286}{2} = R.1430 \quad \text{and} \quad \frac{Ø.015}{2} = R.0075$$

And the final number to be used in the stack-up analysis is R.1430±.0075.

FIGURE 7-7

Inner Boundary of Screw Mounted in Rail = ⌀.2408(LMC Major⌀)−.0140 = ⌀.2268

Outer Boundary of Screw Mounted in Rail = ⌀.250+.014 = ⌀.264

Outer Boundary of Hole in Block = ⌀.286+.015 = ⌀.301

Inner Boundary of Hole in Block = ⌀.276−.005 = ⌀.271

SCREW	BLOCK HOLE
Outer Boundary Mounted Screw = .2640 − Inner Boundary Mounted Screw = .2268 Difference = .0372	Outer Boundary Hole = .301 − Inner Boundary Hole = .271 Difference = .030
1/2 Difference Mounted Screw = .0186	1/2 Difference Hole = .015
Outer Boundary Mounted Screw = .2640 + Inner Boundary Mounted Screw = .2268 Sum = .4908	Outer Boundary Hole = .301 + Inner Boundary Hole = .271 Sum = .572
1/2 Sum of O.B. and I.B. Screw = .2454	1/2 Sum of O.B. and I.B. Hole = .286
1/2 Sum ±1/2 Diff. Screw = .2454±.0186	1/2 Sum ±1/2 Diff. Hole = .286±.015
1/2 of 1/2 Sum ±1/2 of 1/2 Diff. Screw = .1227±.0093	1/2 of 1/2 Sum ±1/2 of 1/2 Diff. Hole = .1430±.0075

The **MIN GAP is calculated** by pushing the block in the cavity to one side (in this case, to the left) so that the right side of the pilot (clearance) hole is seen to touch the right side of the screw. Again, if this is found not to be true, and there is airspace between the mounted screw and the clearance hole, then we may have run the wrong route for our analysis and would have to try another.

We can run another route, for example, to calculate the MIN GAP between the mounted screw and the clearance hole. If we find that route also shows interference is possible, then we can assume that depending on the conditions of the produced part either the edge of the rail may hit the edge of the block, or the edge of the screw mounted in the threaded hole may hit the edge of its clearance hole.

FIGURE 7-8 [Rail Assembly **MIN GAP** Calculation]

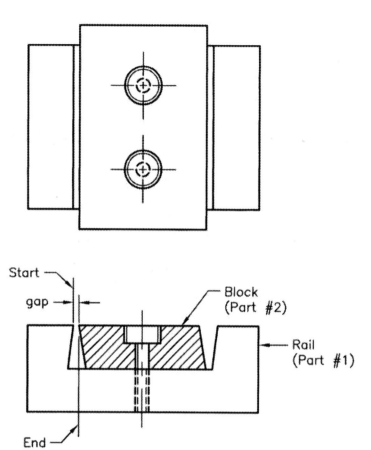

STEP 1:

The loop analysis begins at the left edge of the gap (MIN GAP) and proceeds in the positive (right) direction to the center of the screw.

STEP 2:

Then it continues in the positive direction to the right side of the screw (since the screw is mounted in the slot's threaded hole). We stay on the same part until we have to jump to the mating part.

STEP 3:

We reverse direction, jumping to the mating part and proceed in the negative (left) direction to the center of the clearance hole.

STEP 4:

And, finally, we continue in the negative (left) direction from the center of the clearance pilot hole to the end of the gap loop (the left edge of the block).

FIGURE 7-9 [Rail Assembly **MIN GAP** Calculation]

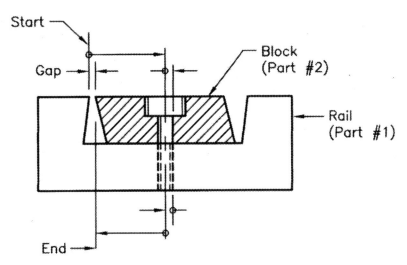

The numbers are inserted into the loop. We have already calculated the numbers for each step in the loop. They are now used in the analysis.

Step 1: +.753±.002 Slot
Step 2: +.1227±.0093 Mounted Screw
Step 3: -.1430±.0075 Clearance Hole
Step 4: -.720±.002 Block

FIGURE 7-10 [Rail Assembly **MIN GAP** Calculation]

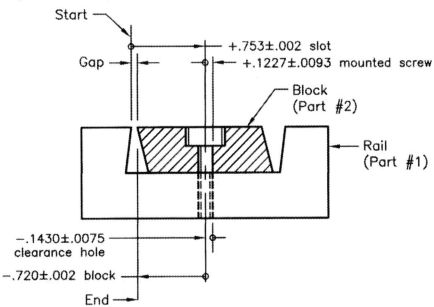

To **calculate the MIN GAP,** we can do one of two things with these numbers. We can add the smallest positives since they represent in Step 1 the smallest slot and the smallest mounted screw. The smallest slot would create the smallest gap. The smallest screw would allow the block to be pushed the most to close the gap. Then we can add the largest negatives since they represent the largest block and the largest clearance pilot hole. The largest block would contribute to the smallest gap and the largest clearance hole would allow the block to be pushed the most (along with the smallest screw) to close the gap.

FIGURE 7-11

So;

MIN GAP

+ _____ (_____ - _____) plus + _____ (_____ - _____) = + _____

− _____ (_____ + _____) plus − _____ (_____ + _____) = − _____

+ _____ plus − _____ = _____ MIN GAP

Or:

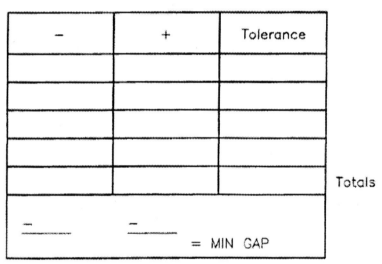

So,

    +.7510 (.753 -.002) = Slot
+ +.1134 (.1227 -.0093) = Mounted Screw
    +.8644 = Numbers for Part #1

and

    -.7220 (.720 +.002) = Block
+ -.1505 (.1430 +.0075) = Clearance Hole
    -.8725 = Numbers for Part #2

These two sets of numbers (for Part #1 and Part #2) are then added to obtain the minimum gap:

    +.8644 = Part #1
+ -.8725 = Part #2
    -.0081 = MIN GAP

This seems to prove the minimum gap, if all features are made in the way described and (this is the key) pushed to close the gap, is actually an **interference** of .0081.

FIGURE 7-12

So:

MIN GAP

+.751 (.753−.002) plus +.1134 (.1227−.0093) = +.8644

−.722 (.720+.002) plus −.1505 (.1430+.0075) = −.8725

+.8644 plus −.8725 = −.0081 MIN GAP

So, the maximum interference is .0081

This method is actually quite interesting in that it can be routed left to right or up and down with lines representing where you start and where you end. We know the MIN GAP to be calculated is between the block and the slot in the cavity. So, when routed left to right, we begin at the left of the GAP and work to the right (as shown in FIGURE 7-10). The left of the GAP is the slot in the cavity, so the first step is left to right. This makes the slot in the cavity a positive step (+.751), the mounted screw a positive step (+.1134), the clearance hole in the block a negative step (-.1505) and the block a negative step (-.722). If we were to rotate the route 90 degrees, the route could be turned into an up and down route. Up and down routes start at the bottom of the GAP and work to the top. It would look like the following illustration.

The CL shown in the middle of the route does not represent the centerline of the assembly, but instead represents the centerline of the feature being worked at that stage of the analysis.

In **Step 1**, the CL represents the centerline of the block.
In **Step 2**, the CL represents the centerline of the clearance hole.
In **Step 3** the CL represents the centerline of the screw that is mounted in the threaded hole.
In **Step 4** the CL represents the centerline of the slot in the cavity.

What appears as one CL on the route in FIGURES 7-13 actually changes to a different CL each time one progresses to the next step in the analysis.

FIGURE 7-13

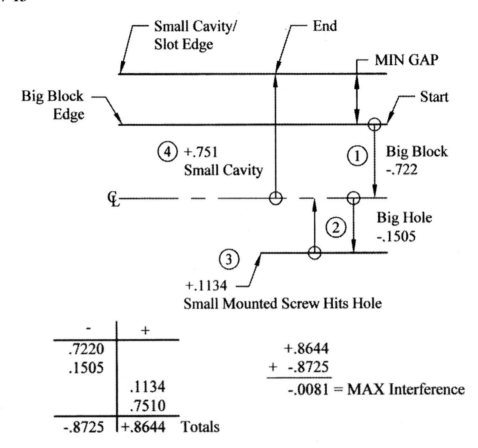

The methodology in FIGURE 7-13 works with only the numbers that would create the MIN GAP (which, in this case, finds a MAX Interference instead).

The methodology suggested in FIGURES 7-10 and 7-11 involves filling in the traditional chart with the loop analysis component numbers:

FIGURE 7-14

MIN GAP

−	+	Tolerance	
	.7530	.0020	
	.1227	.0093	
.1430		.0075	
.7200		.0020	
.8630	.8757	.0208	Totals

.8757 .0127
− .8630 − .0208
.0127   −.0081 = Interference Maximum

114

Each column of numbers is totaled (added). The totals from the negative (minus) and the positive (plus) columns are added. In this case, the positive +.8757 is added to the negative -.8630 and the sum is given as +.0127. To get the minimum gap, the tolerance total of .0208 is subtracted from the .0127. So,

$$+.0127$$
$$- +.0208$$
$$-.0081 \text{ MIN GAP}$$

Since the minimum gap is a negative number, we know there seems to be interference possible of .0081. We know that this interference could occur only if the block was pushed to the left until the mounted screw hit the side of the clearance hole.

Another interesting item to notice is that to get the negative .0081 we subtracted the total tolerance for the features in the route. That would seem to indicate that there is no gap between the block and the slot in the rail. However, if we add the total tolerance of those features we find out that the screws could hit the side of the clearance holes as shown in this calculation (if the block was small and the slot in the rail was large and the tolerances on the screws, threaded holes and clearance holes were used to their maximum advantage). **We find that .0127 plus .0208 equals a maximum gap between the block and the slot in the rail of +.0335 with the block pushed to the left and the screws (mounted in their threaded holes) hitting the right side of the clearance holes.**

Let's examine the logic of pushing the block to the left until the screws that are mounted in the threaded holes hit the right side of the clearance holes and reason out if that is even physically possible with the tolerances of all features in the route subtracted to create the least advantageous situation.

Since this is a situation with only a couple of factors that could contribute to interference on each part, let's first calculate the "worst mating boundaries" of those features.

Features that could interfere during assembly:
1. The inner boundary of the slot in the rail is calculated by taking the MMC of 1.504 and subtracting its perpendicularity tolerance of .002. This equals a "worst mating condition" of 1.502. The outer boundary of the block (that is supposed to fit into the airspace of 1.502) is calculated by adding the MMC of 1.442 and its perpendicularity tolerance of .002. This equals 1.444. If the block has an outer boundary of 1.444 and the slot in the rail has an inner boundary of 1.504, there is no interference between the two.

2. The outer boundary of the screws when they are mounted into the threaded holes is the MMC of the screw which is .250 plus the position tolerance of the threaded hole which is .014. This equals an outer boundary ("worst mating condition") for the mounted screw of .264. The inner boundary of the clearance holes in the block is their MMC of .276 minus their position tolerance of .005. This equals a diameter of .271. If the outer boundary of the mounted screws in the rail's slot is a diameter of .264 and the inner boundary of the clearance holes in the block is a diameter of .271, there is no interference between the two.

Since there are no other features that can interfere, we can deduce that the answer we derived for the MIN GAP of an interference of .0081 is incorrect. Even using the maximum clearance we now know exists between the clearance holes in the block and the screws mounted in the threaded holes in the rail to push the block toward the wall of the slot, **the MIN GAP at its smallest could only be zero.**

In other words, the block touches the slot wall. If the block touches the slot wall, the mounted screws don't likely touch the side of the clearance holes. This means our assumption was wrong and that (using these numbers and this route) we have a clearance between the screw that is mounted in the threaded hole and its clearance hole. The gap (airspace) seems to actually exist between the clearance holes and the screws mounted in the threaded holes.

So, our next step would be to calculate that. If we were to calculate that MIN GAP, it is conceivable (in many situations) that we could also get a negative result, saying that it is possible to use clearances and the pushed conditions to create a zero clearance between the screws that are mounted in the threaded holes and the clearance holes in the block. For example, see FIGURE 7-15.

FIGURE 7-15

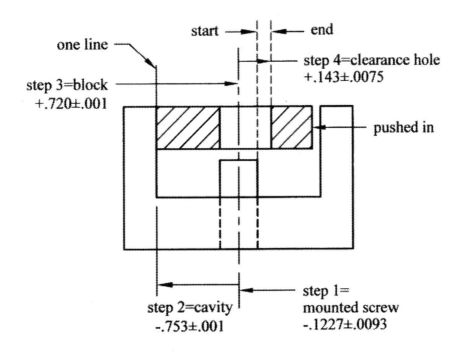

-	+	± tol	item
1. .1227		.0093	mounted screw
2. .753		.001	cavity without perpendicularity
3.	.720	.001	block without perpendicularity
4.	.143	.0075	clearance hole
-.8757	+.863	.0188	totals

```
   +.8630        -.0127                              -.0127
 + -.8757      + -.0188                            + .0188
   ------        ------                              ------
   -.0127        -.0315 = MAX Interference           .0061 = MAX GAP
```

So, as this illustration shows, with the block pushed against the left side of the cavity and the screws already mounted in the threaded holes (with the head of the screws removed), the block might be assembled into the cavity with the clearance holes and the mounted screws clearing by as much as .0061. However, it also shows that the clearance holes and the mounted screws might interfere by as much as much as .0315. It all depends on what sizes the features of each part are produced at and how they are located.

As stated, this particular analysis is done not considering the block and slot perpendicularity as a factor (which for this calculation seems most logical). It is also logical to question the act of trying to assemble these parts with the left side of the block pushed up against the left side of the cavity.

One might ask if proper assembly conditions would use the clearance between all of the mating features to optimize the assembly condition, instead of pushing to one side to force interference. But, in fact, that is what tolerance stack-up analysis often is asked to do. **Still, one should never confuse the question, "Will they assemble if manufacturing uses all of the tolerances available?" with the question, "Can I make them interfere by the way I assemble them?" These parts will assemble, even if all of their tolerances are consumed by manufacturing, if they are assembled optimally.**

To pick up where we left off, discussing what will contact first when one part is pushed to the side: It is most likely the features that will hit first are those features with the tightest fits from part to part. The tightest fit from part to part exists between the clearance holes and the screws (a maximum clearance of a diameter of .0452. as opposed to the block and the cavity which have a maximum clearance of .070-as determined by LMC minus LMC). Even the minimum clearance between the clearance holes and the screws mounted in the threaded holes is tighter than the block in the cavity (.007 vs. .058-as determined by virtual condition minus virtual condition). So, the most likely scenario is that the clearance holes will hit the mounted screws before the block hits the side of the cavity.

Another check of this (once the worst case boundaries have been proven compatible) is to determine the minimum airspace between the mounted screw and its slot (cavity) wall and the maximum wall thickness between the clearance hole and the edge of the block. If there is more airspace than wall thickness, no interference has to occur during assembly. Let's examine the possible interference in that way.

FIGURE 7-16 [Maximum Wall Thickness vs. Minimum Airspace Analysis]

The loop analysis approach assumes the airspace between the screw and the clearance hole will be used to push the parts into the most undesirable assembly conditions. But if we assume the airspace between the screw and the clearance hole is used to push the parts into the most desirable assembly conditions, the analysis proves the parts will assemble.

For example, if the situation is simplified to one threaded hole (with the screw mounted in it) and one clearance hole:

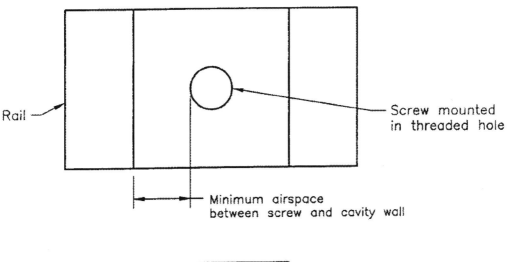

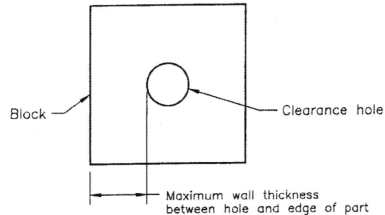

See next page for analysis

There are three basic steps for calculating the MIN AIR between the mounted screw and the slot (cavity) wall. And there are three more steps to calculate the MAX WALL between the surface of the clearance hole and the outside edge of the block.

FIGURE 7-17

Step 1:

$$.250 = \text{MMC of Screw}$$
$$\underline{+ \ .014 = \text{Geo. Tol. of Threaded Hole}}$$
$$\varnothing.264 = \text{Virtual Condition of Mounted Screw}$$

1/2 of ⌀.264 = .132

Step 2:

$$1.504 = \text{MMC Cavity}$$
$$\underline{- \ .002 = \text{Geo. Tol. of Cavity}}$$
$$1.502 = \text{Inner Boundary Cavity}$$

1/2 of 1.502 = .751

Step 3:

$$.751 = 1/2 \text{ Inner Boundary Cavity}$$
$$\underline{- \ .132 = 1/2 \text{ Outer Boundary (\& Virtual Condition) Screw}}$$
$$.619 = \text{MIN AIR between the Screw Surface and the Cavity Wall}$$

Step 4:

$$.276 = \text{MMC Clearance Hole}$$
$$\underline{- \ .005 = \text{Geo. Tol. at MMC}}$$
$$\varnothing.271 = \text{Inner Boundary (Virtual Condition) Clearance Hole}$$

1/2 of ⌀.271 = .1355 = 1/2 Inner Boundary Clearance Hole

Step 5:

$$1.442 = \text{MMC Block}$$
$$\underline{+ \ .002 = \text{Geo. Tol Block}}$$
$$1.444 = \text{Outer Boundary Block}$$

1/2 of 1.444 = .722 = 1/2 of Outer Boundary Block

Step 6:

$$.7220 = 1/2 \text{ Outer Boundary Block}$$
$$\underline{- .1355 = 1/2 \text{ Inner Boundary Hole}}$$
$$.5865 = \text{MAX WALL Thickness bet. Surface of Hole and Outside Edge of Block}$$

Step 7:

$$.6190 = \text{MIN AIR}$$
$$\underline{- .5865 = \text{MAX WALL}}$$
$$.0325 = \text{Clearance between Rail and Block per Side (Minimum Clearance with Parts Adjusted optimally for Assembly)}$$

Therefore, there is no interference necessary when putting these parts together - if the airspace between the screw and the clearance hole is used to adjust the parts to an optimum position for assembly.

The traditional methodology is valid for MAX GAP and MIN GAP calculations but may be misleading. It may arrive at the wrong answer if one is trying to determine from the MIN GAP calculations whether or not the parts will actually fit together if the route chosen assumes the screw touches the side of the clearance hole, but it does not actually touch or even if it does not have to touch and is forced to do so. If the inner boundaries of the holes and the outer boundaries of the shafts of all the mating features are compatible, we can assume the parts are able to fit together if one doesn't force them to interfere.

As stated before, in this case we know the worst case inner boundary of the slot (a hole) is 1.502 and the worst case outer boundary of the block (a shaft) is 1.444. Since the block's worst case assembly condition (1.444) is smaller than the slot's (1.502), they do not interfere.

We also know the worst case outer boundary of the mounted screw is Ø.264 and the worst case inner boundary of the clearance hole is Ø.271. Since the screw mounted in the threaded hole has a worst case assembly condition (Ø.264) that is smaller than its clearance hole's (Ø.271), we know they do not interfere.

Therefore, we know that if the parts are allowed to naturally assemble (are not pushed to extremes by the assembler), they will fit together without interference.

Now, let's use the loop analysis approach to calculate the maximum gap (MAX GAP). To **calculate the maximum gap**, we begin by pushing the block to the right until the left side of the clearance pilot hole hits the left side of the mounted screw.

FIGURE 7-18 [Rail Assembly **MAX GAP** Calculation]

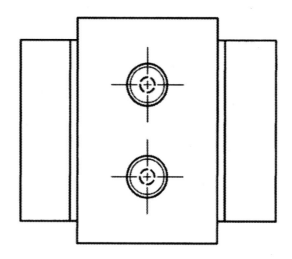

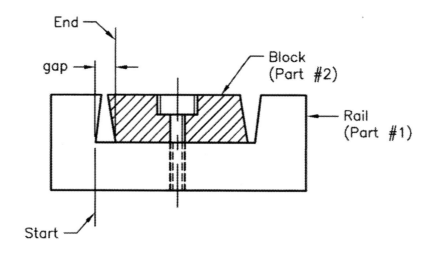

FIGURE 7-19 [Rail Assembly **MAX GAP** Calculation]

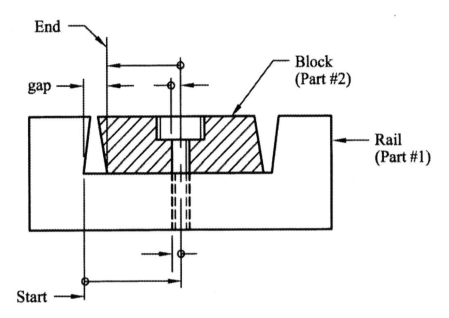

STEP 1:
Again the loop begins at the left edge of the slot and proceeds in the positive (right) direction to the middle of the mounted screw.

STEP 2:
But (unlike the MIN GAP calculation) then we reverse direction, heading to the left (negative) to the middle of the mounted screw (staying on the same part since the screw has become one with the threaded hole and is now referred to as the mounted screw). We go to the left because we can see by the drawing that the left side of the screw is resting against the left side of the clearance hole. We want a continuous stream of pertinent dimensions without jumping over a gap.

STEP 3:
We reverse the loop and go in the positive (right) direction to the center of the clearance pilot hole.

STEP 4:
And, finally, we reverse the loop and go in the negative (left) direction to the end of the gap at the left edge of the block.

The numbers are inserted into the loop. We have already calculated the numbers for each step in the loop. They are now used in the analysis.

Step 1:   +.753±.002 Slot
Step 2:   −.1227±.0093 Mounted Screw in Threaded Hole
Step 3:   +.1430±.0075 Clearance Pilot Hole
Step 4:   −.720±.002 Block

FIGURE 7-20 [Rail Assembly MAX GAP Calculation]

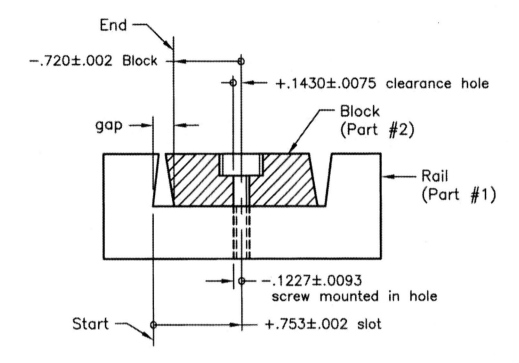

To **calculate the MAX GAP**, we can do one of two things with these numbers. We can add the largest positive values since they represent the largest slot and the largest clearance hole. The largest slot would help create the largest gap and the largest clearance hole would allow the block to be pushed the most to open the gap to its maximum. Then we can add the smallest negatives since they represent the smallest block and the smallest mounted screw. The smallest block would contribute to the maximum gap and the smallest mounted screw would allow the block to be pushed the most (along with the largest clearance hole) to open the gap.

FIGURE 7-21 [MAX GAP]

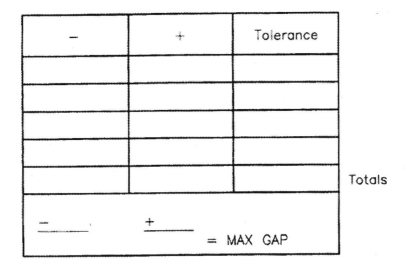

Or:

+ _____ (_____ + _____) plus + _____ (_____ + _____) = + _____

− _____ (_____ − _____) plus − _____ (_____ − _____) = − _____

So; _____ − _____ = _____ MAX GAP

So,

       +.7550 (.753 +.002) = Slot
+ +.1505 (.1430 +.0075) = Clearance Hole
       +.9055

and

       −.7180 (.720 −.002) = Block
+ −.1134 (.1227 −.0093) = Screw
       −.8314

These two sets of numbers are then added to obtain the MAX GAP:

       +.9055
+ −.8314
       +.0741 = MAX GAP

FIGURE 7-22

Or,

+.7550 Largest Slot (O.B.) (.753+.002) plus +.1505 Largest Hole (O.B.) (.1430+.0075) = +.9055 Slot/Hole

-.7180 Smallest Block (I.B.) (.720-.002) plus -.1134 Smallest Screw (I.B.) (.1227-.0093) = -.8314 Block/Screw (Smallest Males)

So,

.9055-.8314 = .0741 MAX GAP

The second method involves filling in the traditional chart with the loop analysis component numbers, as shown in FIGURE 7-23.

FIGURE 7-23 [MAX GAP]

−	+	Tolerance
	.7530	.0020
.1227		.0093
	.1430	.0075
.7200		.0020
.8427	.8960	.0208
.8960 - .8427	.0533 +.0208	
.0533	.0741 = MAX GAP	

Totals (next to the totals row)

Each column of numbers is totaled (added). The totals from the negative (minus) and the positive (plus) columns are added. In this case, positive +.8960 is added to the negative -.8427 and the sum is given as +.0533. To get the maximum gap, the tolerance total of .0208 is added to the .0533.

    .0533
 + .0208
 +.0741 = MAX GAP (**we will find out later in this section that this answer is wrong**)

Unlike the minimum gap wherein the geometric tolerance plays a major role in creating the tightest fits, the maximum gap can be viewed in a couple of ways. One way considers the perpendicularity tolerance of both the cavity and the block and the other does not. The one that does not creates the following illustration. It recognizes that maximum gap, or the most uniform airspace between the two parts, is created when the block and the rail are at their LMC and are perfectly perpendicular.

FIGURE 7-24 [MAX GAP <u>Without</u> Perpendicularity as a Factor]

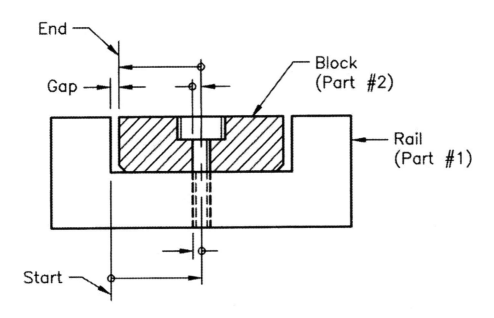

The approach that considers the perpendicularity of the block and the cavity isn't so much considering the uniform maximum airspace between the block and rail, but instead, the MAX GAP between the line elements of the two parts that lean (are out of perpendicularity) away from one another the most.

It seems logical that both are worth knowing and the one that takes into consideration perpendicularity has been shown. But, more realistically, the one that creates the largest uniform open gap is the one that utilizes only least material conditions of the cavity (slot) and the block width. Only perfect perpendicularity would allow for the greatest open uniform gap on either side that could be used to slide the parts around (one within the other). So, let's calculate that more useful maximum gap.

Step 1:
We already know the inner boundary of the mounted screw is Ø.2268 (.2408 LMC - .014 Geo. Tol.), and one-half of that is R.1134.

Step 2:
We know the outer boundary of the clearance hole is Ø.301 (.286 LMC + .015 Geo. Tol.), and one-half of that is R.1505.

Step 3:
The LMC of the cavity is 1.508, and one half of that is R.754.

Step 4:

The LMC of the block is 1.438, and one-half of that is R.719.

It would seem that we now have all the pertinent numbers to do our MAX GAP calculation.

FIGURE 7-25   [MAX GAP <u>without</u> Perpendicularity as a Factor]

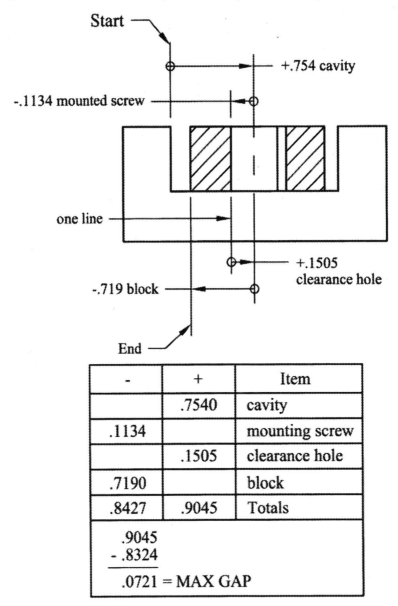

-	+	Item
	.7540	cavity
.1134		mounting screw
	.1505	clearance hole
.7190		block
.8427	.9045	Totals
.9045 − .8324		
.0721 = MAX GAP		

The problem with this answer is that it is easily proven false. There is no way that the clearance hole, the threaded hole and the screw can add to the maximum clearance between the least material conditions of the block and the cavity in the rail. Let's assume for a moment that there was nothing but the block and the rail. The difference between their least material conditions would be the maximum gap possible. Anything else would just act to limit one from touching the other. The LMC of the cavity in the rail is 1.508 (the largest cavity). The LMC of the block is 1.438 (the smallest block). The difference between 1.508 and 1.438 is .070. **So,**

without anything else to limit the MAX GAP, it is .070. Sometimes, the correct approach is the simplest approach.

The problem with the other answer is that it included the threaded hole, screws and clearance holes. Those features should not have been included for the MAX GAP because they were not part of the shortest route. The shortest route excludes features that don't come into contact with one another. Remember, as it said in the beginning of this unit, the correct route for MAX GAP will give you the smallest MAX GAP. This proves that both the .0721 route was not the correct one, and that the .0741 route was not correct. But the real lesson here is to never stop the analysis until you have considered other possible solutions. In Chapter 8 we will examine a variety of other factors, many of which cannot be put into the regular route analysis we have been using. They must be considered afterward. Any analysis result can fool the one performing it, if additional factors or conflicting factors are not considered. See FIGURE 7-26 for the correct route and answer for MAX GAP.

FIGURE 7-26 [MAX GAP <u>without</u> Perpendicularity as a Factor and using the Shortest Route]

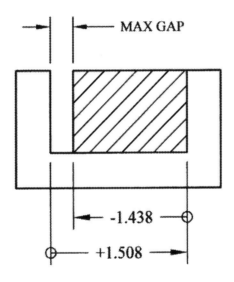

```
  1.508
- 1.438
--------
   .070  = MAX GAP
           shortest route
```

The only other possibility is to calculate the maximum gap, using the shortest route and to include the perpendicularity tolerance on both parts. The problem with this line of thought is that the illustration in FIGURE 7-26 shows that the right side of the block touches the right side of the cavity to create the MAX GAP. If the center planes of the block and the cavity are produced out of perpendicularity by .002 apiece, the surface of the left and right sides of the block would all be out of perpendicularity by .002. Each of the four surfaces would be out of perpendicularity .002. To create the maximum gap possible with this out-of-perpendicularity, this would have the effect of maintaining contact on the bottom right between the cavity and the block. The top right would no longer touch (being separated by the sum of the perpendicularity tolerances .004). The MAX GAP on the bottom left would be unaffected by the out of perpendicularity tolerance. See FIGURE 7-27.

FIGURE 7-27

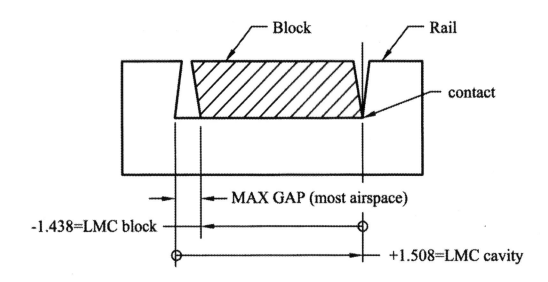

1.508 minus 1.438 equals .070. The interesting and maybe even surprising thing about this illustration is that it shows that the MAX GAP is the same with or without the perpendicularity tolerance. However, since the fit between the block and the cavity is tighter if both are out of perpendicularity as shown in FIGURE 7-27, the perpendicularity tolerance is a factor in calculating the MIN GAP and would make it smaller. That is why it was included in the MIN GAP calculations.

Remember the rules. Take the shortest route and include only those features that are factors in calculating the GAP.

CHAPTER 7
EXERCISES

Exercise 7-1

[Rail Assembly]

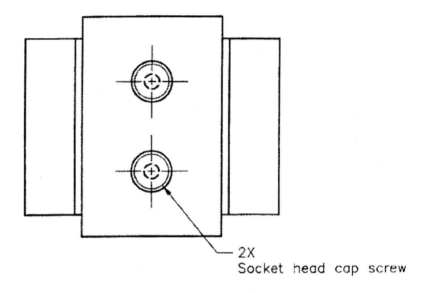

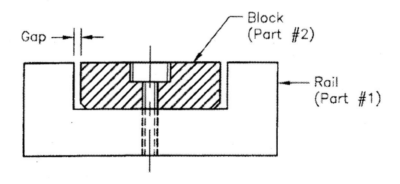

Exercise 7-2

[Rail]

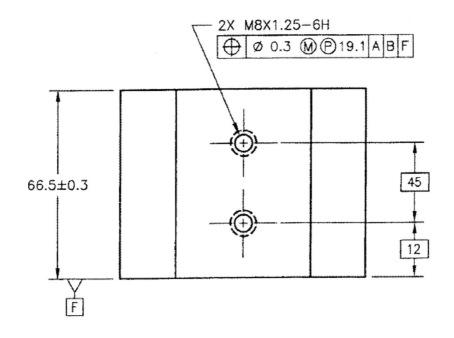

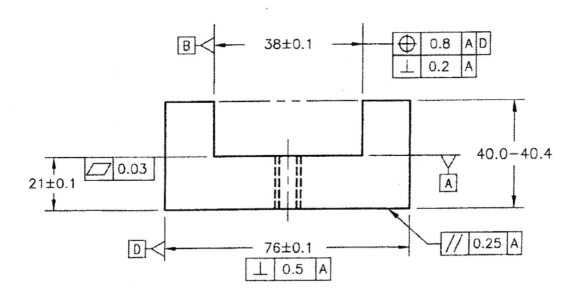

Note: M8 screw: MMC=8; LMC=7.76

Exercise 7-3

[Block]

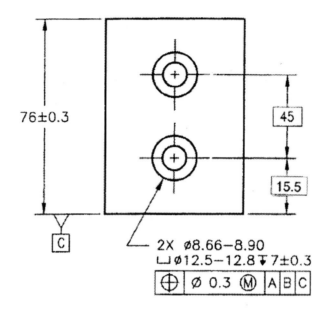

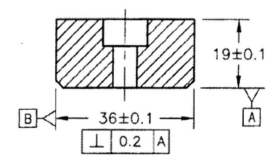

Exercise 7-1 through 7-3, Worksheet #1

[Calculations for Tolerance Stack-Up Analysis for Rail and Block]

Outer Boundary of Slot in Rail = _____

Inner Boundary of Slot in Rail = _____

Inner Boundary of Width of Block = _____

Outer Boundary of Width of Block = _____

Outer Boundary Slot = − Inner Boundary Slot = ──────────────── Difference =	Outer Boundary Block = − Inner Boundary Block = ──────────────── Difference =
1/2 Difference of Slot = _____	1/2 Difference of Block = _____
Outer Boundary Slot = + Inner Boundary Slot = ──────────────── Sum =	Outer Boundary Block = + Inner Boundary Block = ──────────────── Sum =
1/2 Sum of O.B. and I.B. of Slot = _____	1/2 Sum of O.B. and I.B. of Block = _____
1/2 Sum ± 1/2 Diff. of Slot = _____	1/2 Sum ± 1/2 Diff. of Block = _____
1/2 of 1/2 Sum ± 1/2 of 1/2 Diff. Slot = _____	1/2 of 1/2 Sum ± 1/2 of 1/2 Diff. Block = _____

Exercise 7-1 through 7-3, Worksheet #2

Inner Boundary of Screw Mounted in Rail =

Outer Boundary of Screw Mounted in Rail =

Outer Boundary of Hole in Block =

Inner Boundary of Hole in Block =

Outer Boundary Mounted Screw = − Inner Boundary Mounted Screw =	Outer Boundary Hole = − Inner Boundary Hole =
Difference =	Difference =
1/2 Difference Mounted Screw =	1/2 Difference Hole =
Outer Boundary Mounted Screw = + Inner Boundary Mounted Screw =	Outer Boundary Hole = + Inner Boundary Hole =
Sum =	Sum =
1/2 Sum of O.B. and I.B. Screw =	1/2 Sum of O.B. and I.B. Hole =
1/2 Sum ±1/2 Diff. Screw =	1/2 Sum ±1/2 Diff. Hole =
1/2 of 1/2 Sum ±1/2 of 1/2 Diff. Screw =	1/2 of 1/2 Sum ±1/2 of 1/2 Diff. Hole =

Exercise 7-4, Worksheet #1

[Rail Assembly MIN GAP Calculation]

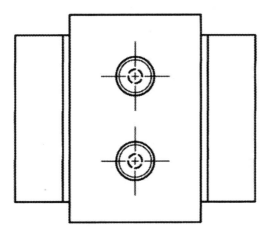

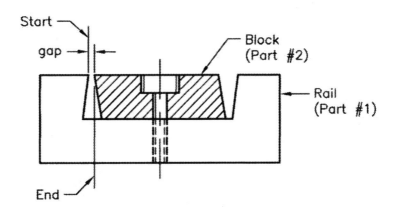

Exercise 7-4, Worksheet #1 (continued)

So;

Min Gap

+ _____ (_____ - _____) plus + _____ (_____ - _____) = + _____

− _____ (_____ + _____) plus − _____ (_____ + _____) = − _____

+ _____ plus − _____ = _____ Min Gap

Or:

MIN GAP

−	+	Tolerance
			Totals

_____   _____   = MIN GAP

Exercise 7-5

The loop analysis approach assumes the airspace between the screw and the clearance hole will be used to push the parts into the most undesirable assembly conditions. But if we assume the airspace between the screw and the clearance hole is used to push the parts into the most desirable assembly conditions, the analysis proves the parts will assemble.

For example, if the situation is simplified to one threaded hole (with the screw mounted in it) and one clearance hole:

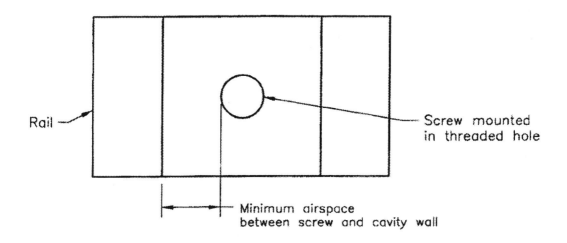

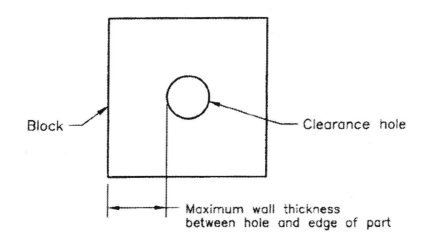

See next page for analysis

Exercise 7-5, Worksheet #1

1. _____ = MMC of Screw
    +_____ = Geo. Tolerance for Threaded Hole
    Ø _____ = Virtual Condition of Screw (while mounted in the threaded hole)
    ÷ 2 = _____ (1/2 the Virtual Condition Screw)

2. _____ = MMC Cavity
    −_____ = Geo. Tolerance for Cavity
    _____ = Inner Boundary of Cavity
    ÷ 2 = _____ (1/2 the Inner Boundary of Cavity)

3. _____ = 1/2 the Inner Boundary of Cavity
    −_____ = 1/2 the Virtual Condition Screw
    _____ = Minimum Airspace (between the screw surface and the cavity wall)

4. _____ = MMC Clearance Hole
    −_____ = Geo. Tolerance for Clearance Hole
    Ø _____ = Virtual Condition of Clearance Hole
    ÷ 2 = _____ (1/2 the Virtual Condition Clearance Hole)

5. _____ = MMC Block
    +_____ = Geo. Tolerance Block
    _____ = Outer Boundary of Block
    ÷ 2 = _____ (1/2 the Outer Boundary Block)

6. _____ = 1/2 the Outer Boundary Block
    −_____ = 1/2 the Virtual Condition Clearance Hole
    _____ = Maximum Wall Thickness (between the surface of the hole to the outside edge of the block)

7. _____ = Minimum Airspace
    −_____ = Maximum Wall Thickness
    _____ = Clearance between Rail and Block per side

Exercise 7-6, Worksheet #1

[Rail Assembly MAX GAP Calculation]

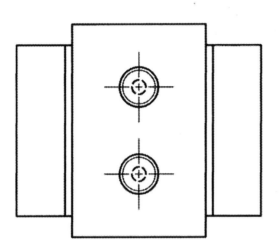

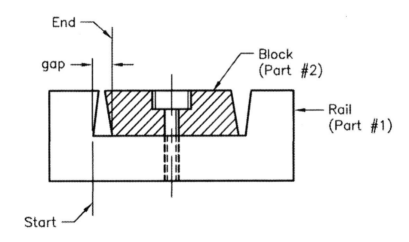

Exercise 7-6, Worksheet #1 (continued)

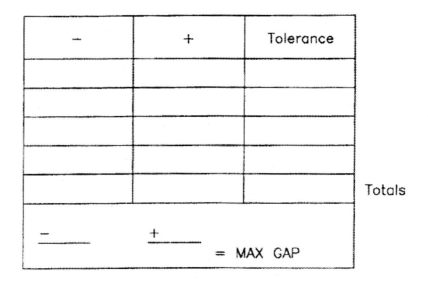

Or:

+ _____ (_____ + _____) plus + _____ (_____ + _____) = + _____

− _____ (_____ − _____) plus − _____ (_____ − _____) = − _____

So; _____ − _____ = _____ MAX GAP

Exercise 7-7, Worksheet #1

Use the following figure to calculate the MAX GAP without including the perpendicularity tolerance allowed for the block and slot in the cavity. Utilize one-half the outer boundary for the clearance hole, one-half the inner boundary for the screw mounted in the threaded hole, one-half the LMC of the cavity and one-half the LMC of the block.

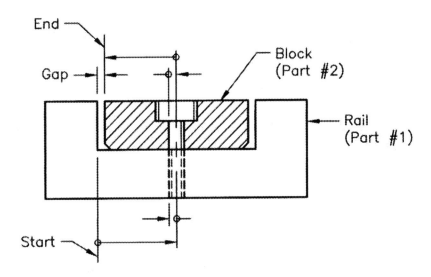

Create a chart below into which the numbers from above will be inserted.

Numbers Chart

Exercise 7-8, Worksheet #1

Utilize the illustration below to either prove or disprove the MAX GAP calculated on the prior worksheet is correct. Use only the LMC of the block and the LMC of the cavity.

Hint: If the answer to this worksheet gives a smaller MAX GAP than on Exercise 7-7, then Exercise 7-7 is incorrect (and the answer to Exercise 7-8 is correct).

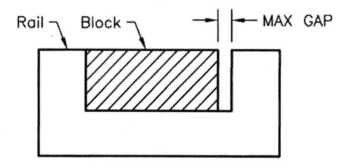

Chapter 8

TOLERANCE STACK-UP ANALYSIS FOR SINGLE PART ANALYSIS

•Lesson Objectives:

In Chapter 8, you will:
- Perform a variety of single-part tolerance stack-up calculations using:
  - profile of a surface
  - two-single segment and composite position tolerancing
  - datum features referenced at Maximum Material Boundary (MMB).
- Analyze envelopes of perfect form and geometric tolerances of perfect orientation at MMC.
- Determine minimum and maximum axial separation.
- Calculate the effects of separate gaging requirements and multiple datum reference frames in accumulating tolerance error.
- Be introduced to trigonometric functions as additional factors in stack-up analysis.
- Determine minimum and maximum wall thicknesses with a variety of approaches.

Chapter 8
SINGLE PART ANALYSIS

FIGURE 8-1

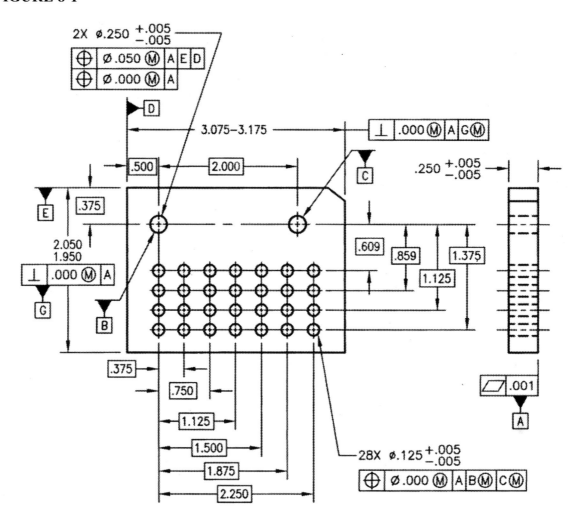

For the single part analysis to be done on FIGURE 8-1, one important factor is to think of the numbers leading away from the end point as eating away material and are, therefore, negative. Numbers leading toward the end point are material and logically are positive. Once the tolerances are considered, if the negatives eat up all the material, the wall breaks out. If not, there is a positive wall.

To begin, we must determine what we are trying to find. In this case, we are going to calculate the minimum wall thickness between the surface of one of the small holes (Ø.120-.130) and the bottom edge of the part shown in the front view. To do this, we must determine all factors that would influence the hole (and the bottom edge of the part). The hole is not measured from the edges of the part. It is measured from the larger Ø.245-.255 hole which is datum feature B.

Datum feature B has a positional tolerance of Ø.050 at MMC to edges of the part (actually datum planes E and D which touch datum features E and D). This tolerance can grow to Ø.060 if datum feature B is produced at Ø.255 (the LMC).

Since the Ø.060 would be the maximum movement allowed to the axis of datum feature B, we will assume the B hole moves the maximum amount (worst case). Wherever the B hole ends up, the small hole is measured from its axis. The small hole has no positional tolerance if it is produced at its MMC (Ø.120), but its positional tolerance can grow to Ø.010 if the hole is produced at a Ø.130. We will assume that it is (worst case). The Ø.130 hole is referenced to datums A, B and C.

Both B & C have been referenced at Maximum Material Boundary (MMB). These allow an additional shift of the pattern of holes away from datum feature axis B (movement of the hole pattern as a group), or an allowed rotation of the hole pattern away from the common datum plane that runs through both B and C. This datum feature shift is only allowed if datum features B and C are produced at something other than their applicable MMB (Ø.245 in this case).

Since C limits pattern rotation and gives us an angle of measurement, while the small hole is measured from the axis of B, C will not be a factor in this problem. It is possible that, in some situations, the rotation of the pattern of holes controlled by datum feature C, has more of an effect than the movement/datum feature shift allowed by B in calculating the MIN and MAX wall thicknesses. In these instances, the rotation must be calculated and compared to the datum feature shift. The largest effect is included in the calculation.

Since B is referenced at MMB (which, in this case, means its applicable virtual condition boundary—and is Ø.245), we should picture B with a Ø.245 gage pin in the hole. Datum feature B actually has two virtual conditions (two MMB's). Its MMB to A, E and D is a Ø.195 (.245 minus .050). Its MMB to A only (and from hole to hole) is Ø.245 (.245 minus .000). Since A is the only datum referenced prior to B at MMB in the position control of the 28 holes, the MMB of Ø.245 if applicable. If the B hole is produced at a diameter of .245, it will hug the gage pin. But, if the hole is produced larger, it may shift aside until it hits the gage pin.

If we understand that the gage pin axis represents the datum axis, we can see that the Ø.130 hole is measured from the datum axis (gage pin). If the datum feature is produced at Ø.255, it can shift a radius of .005 or a diameter of .010. This has the effect of allowing the Ø.130 hole to be farther from the Ø.255 hole by this shift amount (another Ø.010). This thins the minimum wall thickness between the edge of the part and the edge of the Ø.130 hole by the amount of pattern shift.

So, before we begin any loop analysis, we must consider the small hole. It is, for purposes of our analysis, Ø.130. It has a positional tolerance of its own (at that size) of a Ø.010. It is measured from a hole (datum feature B) that has a positional tolerance of Ø.060 (if the hole is produced at the LMC of Ø.255). There is an allowed pattern shift (of the pattern of Ø.130 holes) with datum feature B referenced at MMB. This shift is equal to the difference between the LMC of feature B (Ø.255) and its applicable virtual condition (MMB). Since the only datum feature that precedes

B in the feature control frame (of the Ø.130 holes) is datum feature A. The applicable virtual condition is the hole-to-hole (which includes only A) virtual condition of B, Ø.245 (.245 MMC minus the .000 at MMC position tolerance). Since the difference is Ø.255 minus a Ø.245, the additional pattern shift tolerance is a maximum of Ø.010. So,

$$.130 = \text{LMC of Holes}$$
$$+ \underline{.010 = \text{Geo. Tol. at LMC}}$$
$$\varnothing.140 = \text{Outer Boundary of Hole}$$
$$+ \underline{.060 = \text{Position Tol. of B to Edges of Part}}$$
$$\varnothing.200$$
$$+ \underline{.010 = \text{Pattern Shift (Datum Shift) (B}\text{\textcircled{M}})}$$
$$\varnothing.210 = \text{Hole's Outer Boundary}$$

$$\frac{\varnothing.210}{2} = R.105$$

This radius of .105 is the key number to be used in the loop analysis. The loop analysis could then proceed to calculate the minimum wall thickness by beginning at the bottom edge of the Ø.120-130 hole. It would proceed away from the origin (negative direction) half the hole's effective number of -.105. Then it would go from the center of the hole to the center of datum feature B a negative 1.375 (Basic). It would go to the top edge of the part a negative .375 (Basic). It would then go in the positive direction the inner boundary of the width:

$$1.950 = \text{LMC}$$
$$- \underline{.100 = \text{Geo. Tol at LMC}}$$
$$1.850$$

So,

```
      -.105
   +  -.375                    +1.850 = Material
   + -1.375      and          + -1.855 = Material Eaten
     -1.855                     -.005 = MIN WALL THICKNESS (Breakout)
```

The numbers chart can be created after both inner and outer boundaries have been calculated and the other tolerances affecting the walls figured in. It is capable of calculating both the minimum and the maximum wall thicknesses. The Outer Boundary Hole Calculation has already been calculated at .130+.010 = Ø.140. Then the position tolerance of a maximum of Ø.060 was added as the position tolerance of B (since the 28 holes are measured from B). And last, the Ø.010 was added for possible pattern shift (datum feature shift) because B was referenced by the 28 holes at MMB. The total was .210.

Now the inner boundary is calculated by the following procedure:

$$.120 = \text{MMC Hole (28 hole pattern)}$$
$$- \underline{.000 = \text{Geo. Tol. applicable at MMC}}$$
$$\varnothing.120 = \text{Inner Boundary of Hole}$$
$$- \underline{.060 = \text{Position Tol. of B}}$$
$$\varnothing.060 = \text{Inner Boundary minus Position of B}$$
$$- \underline{.010 = \text{Pattern Shift (B referenced at MMB)}}$$
$$\varnothing.050 = \text{Inner Boundary of 28 Holes (minus other pertinent tolerances)}$$

Now we subtract the two boundaries, and then divide that figure by 2:

Ø.210 = Outer Boundary Hole
- Ø.050 = Inner Boundary Hole
Ø.160 = Difference

$$\frac{Ø.160}{2} = Ø.080$$

Then we add the boundaries and divide by 2. So,

Ø.210
+ Ø.050
Ø.260

$$\frac{Ø.260}{2} = Ø.130$$

We express the diameter as an equal bilateral toleranced dimension of Ø.130±.080.

Now we express it as a radius by dividing by 2 again.

$$\frac{.130}{2} = .065 \quad\quad \text{and} \quad\quad \frac{.080}{2} = .040$$

So, the radial expression is: R.065±.040

The numbers chart is easily created from this point. The numbers leading us from the bottom of the small hole to the top of the part are .065, 1.375 and .375. Since these all eat into the minimum wall thickness and go away from the end point in the wall thickness calculation, they will be negative numbers. The dimension being eaten into by the negative numbers, and to be expressed as a positive number, is: 2.050 (2.050 MMC + .000 Geo. Tol. at MMC) to 1.850 (1.950 LMC - .100 Geo. Tol. at LMC). So the difference of .200 is expressed as ±.100, and the mean is 1.950. So we get 1.950±.100. We now insert these numbers into our chart.

FIGURE 8-2

–	+	±Tol	
	1.950	.100	Part
.375		.000	Basic
1.375		.000	Basic
.065		.040	Hole
1.815	1.950	.140	Totals

We add the totals:
1.950 = Positive Total
+ -1.815 = Negative Total
.135

Then we subtract the tolerance to get the minimum wall thickness. So,

    .135=Mean Wall Thickness
  - .140=Sum of Tolerances
  - .005 = MIN WALL THICKNESS (Breakout)

From these same numbers we can calculate the maximum wall thickness. For example:

  +.135= Mean Wall Thickness
  +.140=Sum of Tolerances
  .275=MAX WALL THICKNESS

FIGURE 8-3 [Loop Analysis Chart 1]

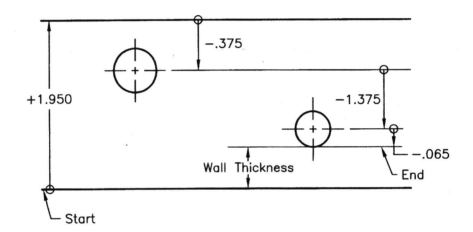

−	+	±Tol	
	1.950	.100	Part
.375		.000	Basic
1.375		.000	Basic
.065		.040	Hole
1.815	1.950	.140	Totals

    1.950   .135   .135
  - 1.815   - .140   + .140
    .135   -.005=MIN WALL   .275=MAX WALL
                    (Breakout)

Once the key number of R.065±.040 for the small hole has been found, any wall thickness between the small hole and an edge close to the small hole can be found. For example, the wall on the right edge of the part can be found with the following numbers chart. First we identify the negative numbers as the .065, the 2.250 and the .500. This leads us to the left edge of the part. To get back to the right edge, we use the 3.075-3.175 dimension. But we must calculate the inner and outer boundaries.

Outer Boundary = 3.175+.000 (Geo. Tol.) = 3.175
Inner Boundary = 3.075-.100 (Geo. Tol. at MMC) = 2.975

First we add and then divide by 2 to get the mean dimension.

```
    3.175 = Outer Boundary
 + 2.975 = Inner Boundary        and        6.150 = 3.075
   6.150                                      2
```

Then we subtract and divide by 2 to get the plus and minus tolerance.

```
    3.175
  - 2.975                         and        .200 = .100
     .200                                      2
```

So, the dimension expressed with an equal bilateral tolerance is 3.075±.100. It is the material we will be eating away at with the negative numbers, so it is expressed as a positive number. It leads us toward the end point.

FIGURE 8-4

−	+	±Tol	
.065		.040	Hole
2.250		.000	Basic
.500		.000	Basic
	3.075	.100	Part
2.815	3.075	.140	Totals

We add the totals:
```
   +3.075 = Positive Total
 + -2.815 = Negative Total
   + .260 = Mean Wall Thickness
```

Then we subtract the tolerance to get the minimum wall thickness between the right side of the small hole and the right side of the part. This calculation does not factor in the effects of the part periphery's perpendicularity tolerances. But, these can always be factored in at the end of any problem. These and other additional factors that may or may not affect the final answers are investigated in subsequent problems in this chapter.
So,
```
    .260 = Mean Wall Thickness
  - .140 = Sum of Tolerances
    .120 = MIN WALL THICKNESS
```

And, again, to calculate the maximum wall thickness, we add the tolerance.
```
    .260 = Mean Wall Thickness
  + .140 = Sum of Tolerances
   +.400 = MAX WALL THICKNESS
```

FIGURE 8-5

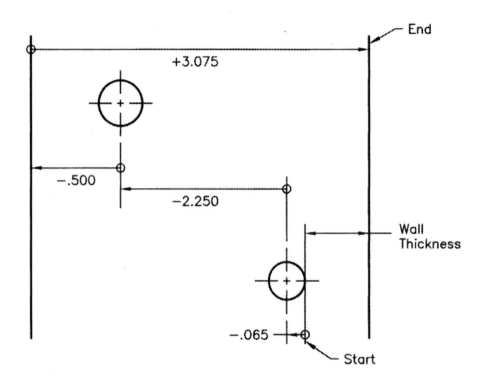

–	+	±Tol	
.065		.040	Hole
2.250		.000	Basic
.500		.000	Basic
	3.075	.100	Part
2.815	3.075	.140	Totals

```
 3.075       .260              .260
-2.815      -.140             +.140
 -----      --------------    --------------
  .260      .120=MIN WALL     .400=MAX WALL
```

And to get the maximum distance allowed between the right-hand edge of the part and the axis of the small hole, we add the .065 back in and get .400+.065 = .465. Of course, we keep the tolerance of plus or minus .040 as a factor to get the .465 because the tolerance describes movement allowed the hole and remains a factor. Remember, the .065 is simply the radius of the hole size at LMC. In other words, one-half of Ø.130 (the largest allowed hole diameter, its LMC) is R.065. Everything else is a form of movement and, therefore, not a factor. All the R.065 allows us to do is get from the edge of the physical hole to the center of the hole. All other tolerances of movement have already been factored in.

By that same process, we can add the .065 to the .220 minimum wall thickness to get the minimum distance allowed between the right-hand edge of the part and the axis of the hole.

$$.120 = \text{MIN WALL THICKNESS}$$
$$+ .065 = \text{Radius of the Hole LMC}$$
$$.185 = \text{Minimum Distance from Hole Center to Right-Hand Edge of Part}$$

There are other ways to do these analyses, but this unit shows one of the easiest, shortest and most logical. These analyses have been the most difficult relationships this part offers. The same approach could be used to calculate some of the simpler relationships.

Hole to hole relationships, for example, are very easy compared to the previous analyses. To use the loop analysis approach, we could simply recognize that the 2.000 basic dimension (from hole to hole) is toleranced by the zero at MMC tolerance. This being the case, we calculate inner and outer boundaries as follows:

Step 1:

$$.245 = \text{MMC Hole}$$
$$- .000 = \text{Geo. Tol. at MMC}$$
$$\varnothing.245 = \text{Inner Boundary Hole}$$

Step 2:

$$.255 = \text{LMC Hole}$$
$$+ .010 = \text{Geo. Tol. at LMC}$$
$$\varnothing.265 = \text{Outer Boundary Hole}$$

Step 3:
Add the boundaries:

$$.265$$
$$+ .245$$
$$\varnothing.510$$

Step 4:
Divide by 2 to get the mean hole:

$$\frac{\varnothing.510}{2} = \varnothing.255$$

Step 5:
Subtract the boundaries to get the tolerance:

$$.265$$
$$- .245$$
$$.020$$

Step 6:
Divide by 2 to express the tolerance as plus or minus:

$$\frac{.020}{2} = \pm.010$$

Step 7:
Take the mean hole with its plus or minus tolerance, which is Ø.260±.010 and divide both by 2 to express each hole as a radius:

$$\frac{\varnothing.255}{2} = R.1275$$

and

$$\frac{.010}{2} = R\pm.005$$

Step 8:
Combine the radial dimension and tolerance:

R.1275±.005

Now we can do a loop analysis.

FIGURE 8-6

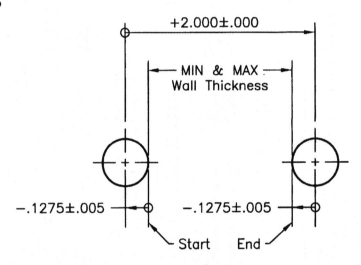

−	+	±Tol	
.1275		.005	
	2.000	.000	
.1275		.005	
.255	2.000	.010	Totals

Add Totals	Subtract Tolerance	Add Tolerance
2.000	1.745	1.745
+ -.255	- .010	+ .010
1.745	1.735=MIN WALL	1.755=MAX WALL

To get axis distances, add back in half the hole LMC diameters:

$$\frac{\varnothing.255}{2} = \begin{array}{l} .1275 \\ \times\ 2 \\ \hline .255 \end{array} = \begin{array}{l} \text{Hole Radius} \\ = \text{2 Holes} \\ = \text{Sum of 2-Hole Radii} \end{array}$$

$$\begin{array}{l} 1.730 \\ +\ .260 \\ \hline 1.990 = \text{MIN Separation} \\ \text{Allowed Between Hole Axes} \end{array} \qquad \begin{array}{l} 1.750 \\ +\ .260 \\ \hline 2.010 = \text{MAX Separation} \\ \text{Allowed between Hole Axes} \end{array}$$

This same technique can be used to calculate the minimum and maximum wall thickness to either datum feature E or D. In this case, though, we must consider that the geometric tolerance that applies is the only one that referenced E and D. This is the positional tolerance of .050 at MMC. The tolerance would grow to .060 at LMC. So, to begin:

Step 1:
Calculate the inner and outer boundaries of the hole called B:

.195 = Inner Boundary Hole (.245 -.050 MMC - Geo. Tol at MMC)

.315 = Outer Boundary Hole (.255 + .060 LMC + Geo. Tol. at LMC)

Step 2:
Find the mean dimension:

$$\begin{array}{l} .315 = \text{Outer Boundary Hole} \\ +\ .195 = \text{Inner Boundary Hole} \\ \hline .510 = \text{Sum} \end{array} \qquad \frac{.510}{2} = \varnothing.255 = \text{Mean Diameter Boundary}$$

Step 3:
Find the ± tolerance:

$$\begin{array}{l} .315 = \text{Outer Boundary Hole} \\ -\ .195 = \text{Inner Boundary Hole} \\ \hline .120 = \text{Tol.} \end{array} \qquad \frac{.120}{2} = .060 = \pm\ \text{Tol.}$$

Step 4:
Convert both to radii:

$$\frac{.255}{2} = R.1275$$

and $\qquad \dfrac{.060}{2} = R.030$

Step 5:
So:

.1275±.0300

Now we can do a loop analysis.

FIGURE 8-7

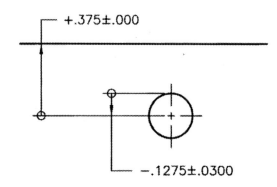

So, to datum plane E, we can say:
```
   +.3750        .2475
+ -.1275       - .0300
   .2475        .2175=MIN WALL to the datum plane E
```

But datum surface E may be out-of-flatness by the overall size tolerance, so a more final answer would be:
```
   2.050
 - 1.950
    .100
```
and
```
    .2175
  - .1000
    .1175=MIN WALL to the lowest point on datum surface E
```

FIGURE 8-8 [Explanation of FIGURE 8-7]

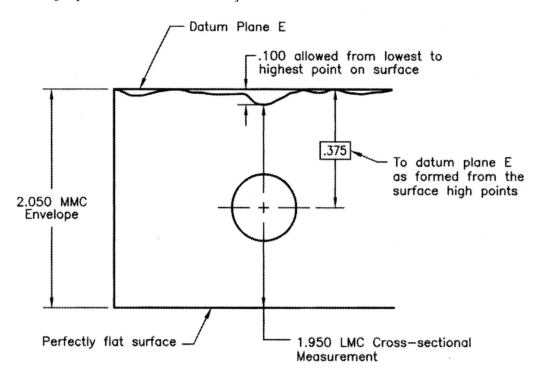

156

FIGURE 8-9

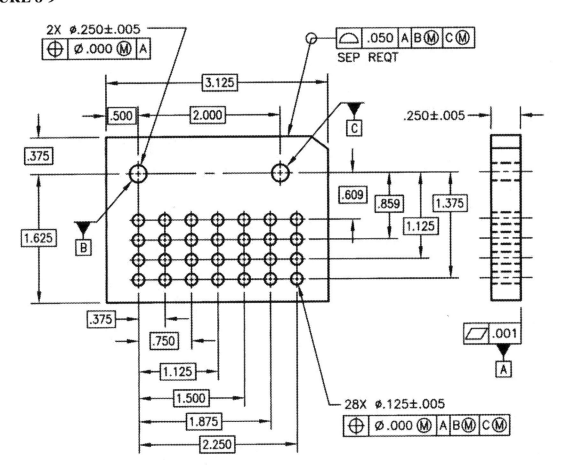

This example is interesting in the variety of ways it will introduce to analyze the tolerances. To begin, there is a 2-hole pattern of holes with a position control as follows:

2X ⌀.250±.005
| ⌖ | ⌀.000 Ⓜ | A |

This is a control of the holes to each other and to datum plane A. The relationship to datum A is a simple perpendicularity control and will not be elaborated on here. The more interesting part of this control is the hole-to-hole tolerance. This effectively tolerances the two holes in their movements away from the 2.000 basic dimension. This analysis can be thought of in terms of tolerance zones or in terms of virtual condition boundaries.

First let's examine them in terms of tolerance zones. The zones exist on either end of the 2.000 basic dimension. The zones are zero at maximum material condition (⌀.245), but as each hole grows from that size, the positional tolerance grows. Since each hole can grow to ⌀.255 (the Least Material Condition), each hole's positional tolerance can grow to a maximum of ⌀.010. An illustration of this is shown on the next page.

FIGURE 8-10

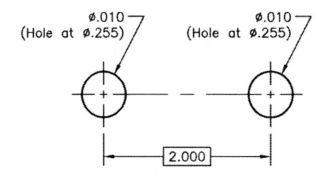

This means the axes of the holes may be a maximum distance apart of 2.010 and a minimum distance apart of 1.990.

FIGURE 8-11

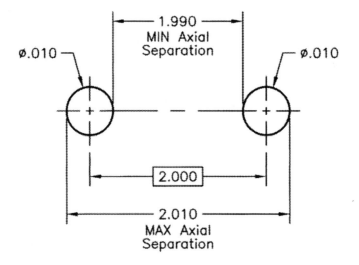

Another way to consider this is through the virtual condition boundaries. The virtual condition of each hole is Ø.245 (.245 MMC minus .000 geometric tolerance). Since functional gage pins would be designed at the virtual condition size, we could consider that a Ø.245 gage pin is in each hole and these gage pins (representing the virtual condition boundaries) are 2.000 apart from center to center (axis to axis). If the holes are produced at Ø.245, each would (in order to comply with their positional tolerance requirement) hug its gage pin.

FIGURE 8-12

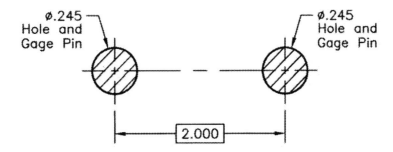

But if the holes were produced larger, they could be out of position in any radial direction half of their growth. If each hole was produced at a Ø.255, the result could be viewed as follows.

FIGURE 8-13

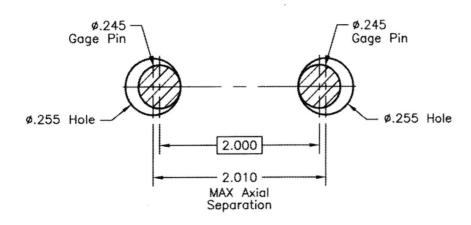

Or:

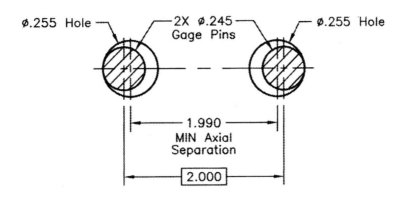

The profile control relates the outside of the part to the two holes. Let's consider the relationship between the left upper corner (in the front view) of the part and the hole which becomes datum feature B. Their basic relationship is shown in FIGURE 8-14.

FIGURE 8-14

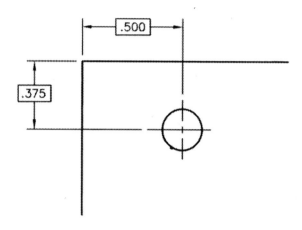

This is toleranced by a fairly complex profile control of:

To begin, we can consider the profile (edges of the part--all around) can grow or shrink by plus or minus .025 per surface. This could be thought of affecting the two basic dimensions by ±.025. So, just considering the profile tolerance of .050, we could see that the distance between these actually produced edges of the part (profile) and the axis of the holes are .500±.025 (.475 to .525) and .375±.025 (.350 to .400).

The overall dimensions of 2.000 (1.625 + .375) and 3.125 are affected by ±.025 per surface. This means the overall size of the part is 2.000±.050 x 3.125±.050. The datum features referenced have no effect on the size of the part (only on the angle and location of the part surfaces).

However, the fact that datum feature B is referenced at MMB (maximum material boundary) does have an effect on the .500 and .375 dimension beyond the .050 of profile tolerance. Illustrations of the extremes are as follows:

FIGURE 8-15

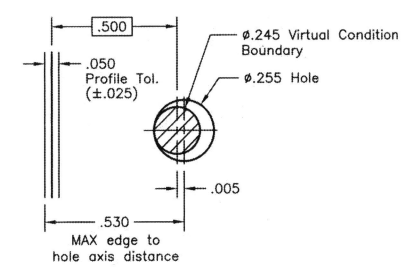

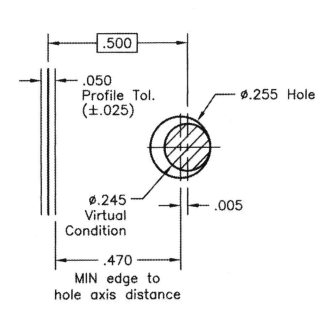

Likewise, the .375 dimension is affected by a maximum tolerance of ±.030, so its limits are a maximum edge to hole axis distance of .375 + .030 = .405 and a minimum of .375 - .030 = .345.

The C datum feature referenced at MMB (maximum material boundary) does not add to this tolerance. In fact, the job of the C datum is to further limit the degrees of part freedom. The common plane between the B and C datum axes (think about it as the common plane between the gage pins of Ø.245 located at 2.000 apart) give us an angle to measure the profile along. This C datum is to orient the planes crossing at the axis of B. To determine the minimum and maximum wall thicknesses (from the edge of the part to the edge of the hole), we have only to use half the hole diameter.

FIGURE 8-16

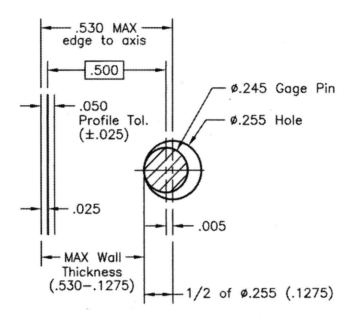

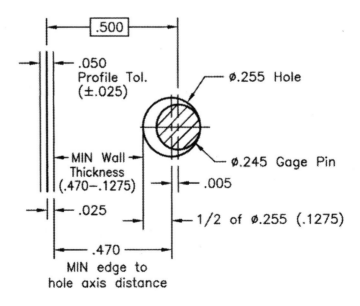

        .5000 = Basic Dimension
- .0050 = 1/2 of .010 Tolerance from B at MMB

        .4950
- .1275 = 1/2 of Ø.255 (Hole's LMC)

        .3675
- .0250 = 1/2 Profile Tolerance

        .3425 = MIN WALL THICKNESS

Or . . .

        .4700 - .1275 = .3425 MIN WALL to left edge of part

and

        .5300 = MAX Distance from hole's axis to left edge of part
- .1275 = ½ of Ø.255 (Hole's LMC)

        .4025 = MAX WALL Thickness to left edge of part

and

        .3450 = (.375 minus .025 profile tolerance minus .005 datum feature shift)
- .1275 = ½ of Ø.255 (Hole's LMC)
        .2175 = MIN WALL Thickness to top edge of part

and

        .4050 = (.375 plus .025 profile tolerance plus .005 datum feature shift)
- .1275 = ½ of Ø.255 (Hole's LMC)
        .2775 = MAX WALL Thickness to top edge of part

Loop analysis charts could be done as shown in FIGURE 8-17.

FIGURE 8-17

—	+	±Tol	
	.500	.025	
.1275		.005	
−.1275	+.500	.030	Totals

Add Totals	Subtract Tolerance	Add Tolerance
+.500	.3725	.3725
+ −.1275	− .0300	+ .0300
+.3725	.3425=MIN WALL	.4025=MAX WALL

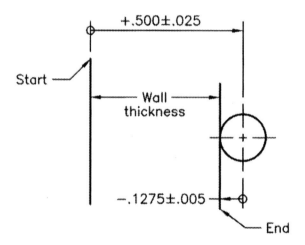

Even more involved a problem would be to consider the smaller 28 holes. These holes have an outer and inner boundary calculated by the following method.

FIGURE 8-18 [Calculations]

```
  .120 = MMC                        .130 = LMC
- .000 = Geo. Tol. at MMC         + .010 = Geo. Tol. at LMC
 Ø.120 = Inner Boundary            Ø.140 = Outer Boundary
- .010 = Pattern Shift (B Ⓜ)      + .010 = Pattern Shift (B Ⓜ)
 Ø.110 = Inner Boundary with       Ø.150 = Outer Boundary with
         Pattern Shift                     Pattern Shift
```

```
  Ø.150
+ Ø.110
  Ø.260 = Sum
```

Ø.260 ÷ 2 = Ø.130 = Mean Dimension

```
  Ø.150
- Ø.110
  Ø.040 = Difference
```

Ø.040 ÷ 2 = Ø.020 = Mean Plus and Minus Tolerance

So the mean dimension and tolerance is Ø.130±.020. Expressed as a radius, it is R.065±.010.

The loop analysis can be done with these numbers but must consider that the profile tolerance on the right edge is ±.025, with an additional datum feature shift (due to B referenced at Maximum Material Boundary) a **separate requirement** from the 28 holes of another ±.005 (since the edge could shift one way and the 28 holes the opposite way).

FIGURE 8-19

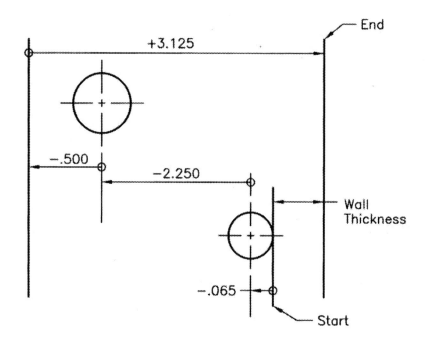

−	+	±Tol	
.065		.010	Hole
2.250		.000	Basic
.500		.000	Basic
	3.125	.030	Part (.025+.005)
2.815	3.125	.040	Totals

Add Totals	Subtract Tolerance	Add Tolerance
3.125	.310	.310
+ −2.815	− .040	+ .040
.310	.270=MIN WALL	.350=MAX WALL

Technically, this route is incorrect. It isn't the shortest route. The .500 dimension is unnecessary to include in this analysis. It is merely a vehicle to subtract from the overall 3.125 basic dimension. This route only works if the profile tolerance on the left edge of the part is excluded from the analysis. In this way, unnecessary accumulated tolerance error is not added to the problem. **It is much better to use the shortest route shown in FIGURE 8-20.**

Even though they are separate requirements and may shift in opposite directions, both the right edge of the part and the 28 holes are located from datum axis B. The basic distance from B to each is given in the loop analysis below.

FIGURE 8-20

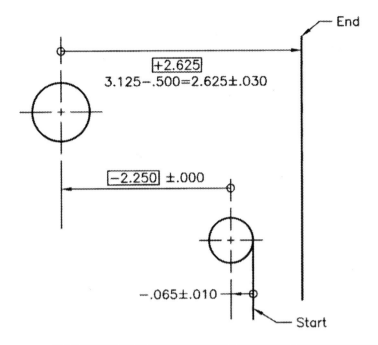

Add Negatives	Add Totals	Subtract Tol.	Add Tol.
−2.250 + −.065 −2.315	+2.625 + −2.315 .310	.310 − .040 .270=MIN	.310 +.040 .350=MAX

To check our calculations, we can employ an alternate approach to figure the minimum wall thickness as follows:

```
 .130=LMC Hole              2.625         .300
+.010=G.T. at LMC          -2.250        -.005=right edge shift
Ø.140=O.B. Hole              .375         .295
+.010=Pattern shift                      -.025=profile tol. radius
Ø.150=O.B. Hole w/shift      .375         .270=MIN Wall
                            - .075
                             .300

Ø.150/2=.075=1/2 O.B. Hole w/shift
```

To get back to the middle of the hole, we add the radius of the LMC only (not including position tolerance). One-half of Ø.130 is .065.
So,
.270
+ .065
.335 = MIN Distance from Right Edge to Hole Axis

Since the MIN WALL proved correct when the alternate approach was employed, we can assume the MAX WALL was also correct. And to get the maximum distance from the right edge of the part to the center of the hole, we can add the same hole radius (.065) from the previously calculated .350 MAX WALL THICKNESS.

.350
+ .065
.415 Maximum Distance Allowed (between the right hand edge of the part and the axis of the Ø.130 hole closest to it if all holes are produced at LMC)

Employing an alternate approach proves it to be correct. For example, take all pertinent basic dimensions and subtract them to get the basic distance from the axis of the small hole to the right edge of the part: 3.125 - .500 - 2.250 = .375. Now, to get the maximum distance, add all applicable tolerances:

.375 = Basic Distance from Axis to Right Edge of Part
+ .025 = Half the Profile Tol. on Right Edge (for growth)
+ .005 = Half the Shift from B of Right Edge (Profile to A and **B at MMB**)
+ .005 = Half the Position Tol. of Ø.130 Hole
+ .005 = Half the Shift of Ø.130 Hole Axis from B (Position to A and **B at MMB**)
[You get this again because the profile and the 28 holes are a Separate Requirement.]

.415 = MAX Distance from Right Edge of Part to Ø.130 Hole Axis

Then, to get the MAX WALL THICKNESS from the right edge of the hole to the right edge of the part, we subtract half the hole size (Ø.130 ÷ 2 = R.065).
So,
.415
- .065
.350 MAX WALL THICKNESS

CHAPTER 8
EXERCISES

Exercise 8-1, Problems #1 and #2

PROBLEM #1
Calculate the MIN and MAX WALLS and axis distances from the left edge of the part to the ⌀.125±.005 hole closest to it.

PROBLEM #2
Calculate the MIN and MAX WALLS from the bottom edge of the part to the ⌀.125±.005 hole closest to it.

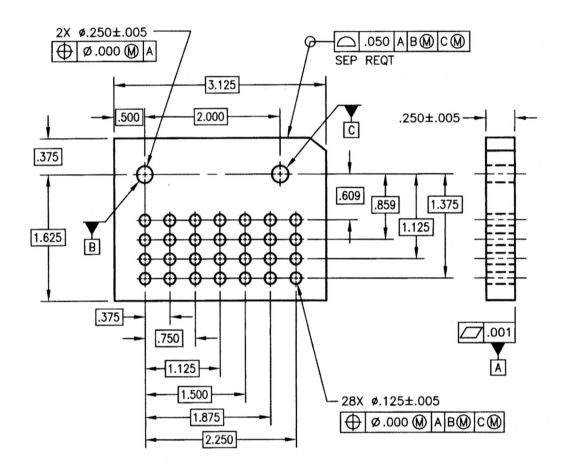

Exercise 8-2, Problems #1, #2 and #3

PROBLEM #1
What is the minimum and maximum wall thickness between the B and C holes if they are produced at LMC?

PROBLEM #2
What is the minimum and maximum wall thickness from the B hole to the top edge of the part shown in the front view?

PROBLEM #3
What is the minimum and maximum wall thickness from the lower right ⌀4.0-4.3 hole to the bottom edge of the part in the front view?

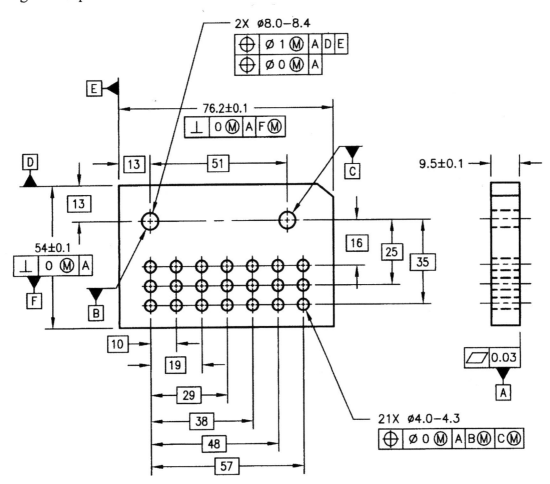

Exercise 8-3, Problems #1, #2 and #3

PROBLEM #1
What is the minimum and maximum wall thickness between the B and C holes if they are produced at LMC?

PROBLEM #2
What is the minimum and maximum wall thickness from the B hole to the top edge of the part shown in the front view?

PROBLEM #3
What is the minimum and maximum wall thickness from the lower right ⌀4.0-4.3 hole to the right edge and the bottom edge of the part in the front view?

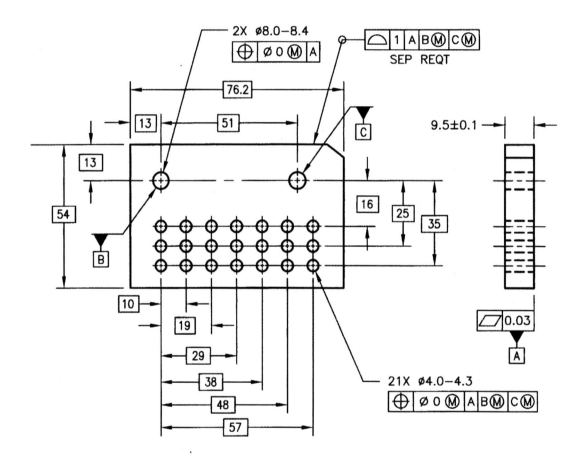

Exercise 8-4, Problems #1

PROBLEM #1
Calculate the minimum and maximum wall thickness between the surface of one of the holes and the outside diameter of the part.

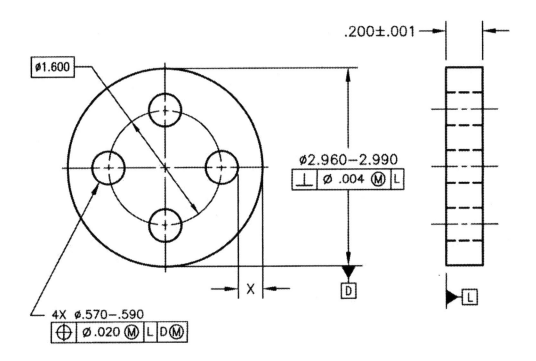

Exercise 8-5, Problems #1

PROBLEM #1
What is the minimum wall thickness between one of the holes nearest the top of the part in the front view and datum feature C?

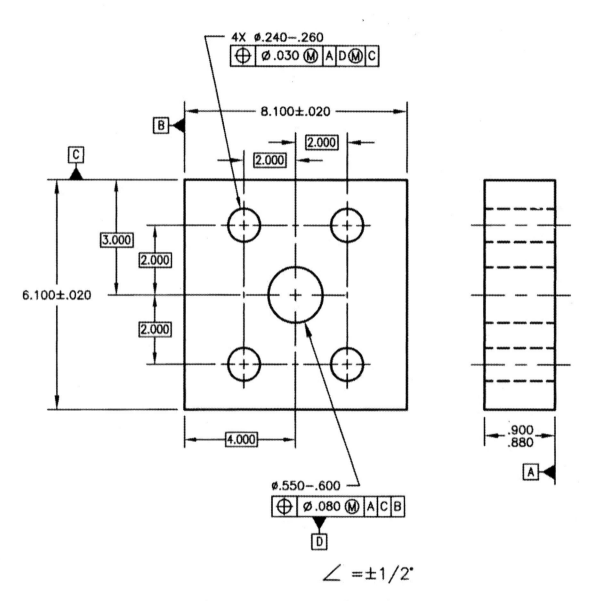

Exercise 8-6, Problems #1

PROBLEM #1: What is the minimum wall thickness of the hole to the right edge of the part?

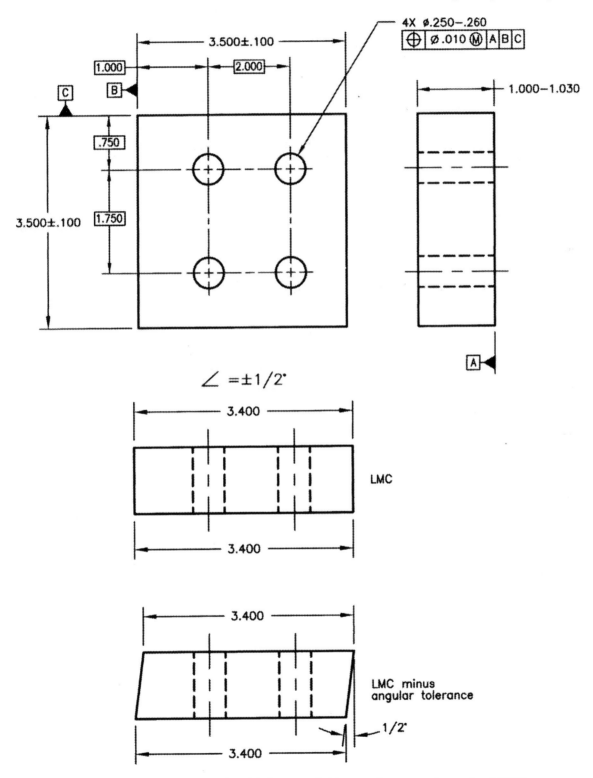

One-half of a degree over the thickness of this part is approximately .00873.

Exercise 8-7, Problems #1

PROBLEM #1: Calculate the minimum wall thickness of one of four holes in the pattern to the outside diameter of the part.

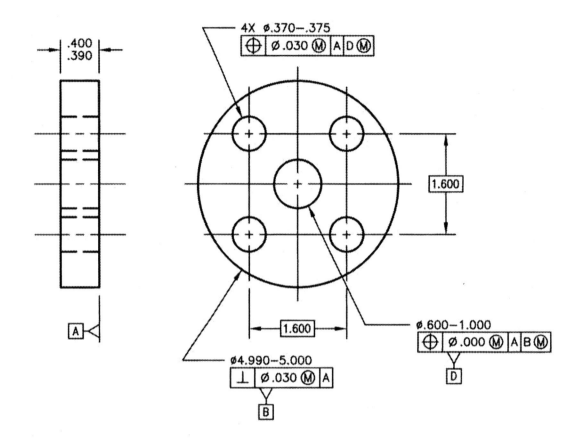

Exercise 8-8, Problems #1

PROBLEM #1: In the right side view of the detail drawing, what are the minimum and maximum wall thicknesses between the edge of the hole and the left edge of the part?

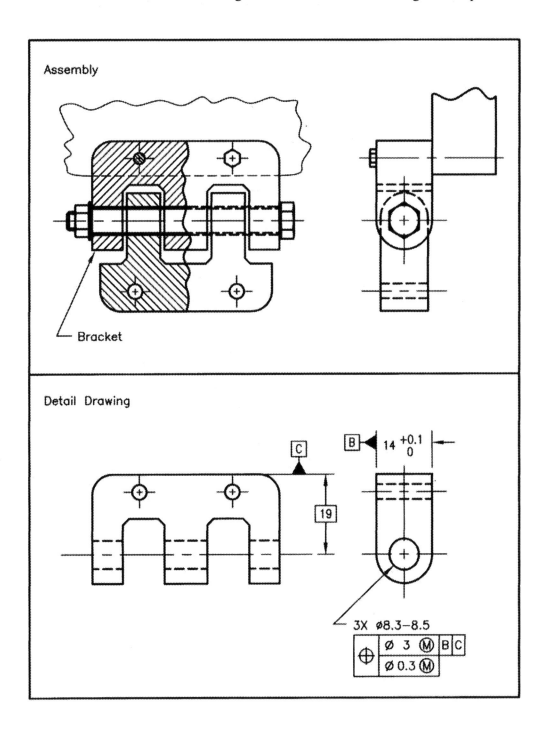

Chapter 9

TOLERANCE STACK-UP ANALYSIS FOR A FIVE-PART ASSEMBLY

• Lesson Objectives:

In Chapter 9, you will:
- Perform a tolerance stack-up analysis on a complex five-part rotating assembly with a wide variety of geometric tolerances, such as: position; perpendicularity; parallelism; profile; flatness; projected tolerance zones; runout; total runout; concentricity; positional coaxiality
- Practice simplifying a complex situation.
- Learn to calculate part-to-part analysis from two parts to an infinite number of parts.
- Determine assembly housing requirements.
- Calculate radial clearance and interference.

FIGURE 9-1 [Five-Part Assembly]

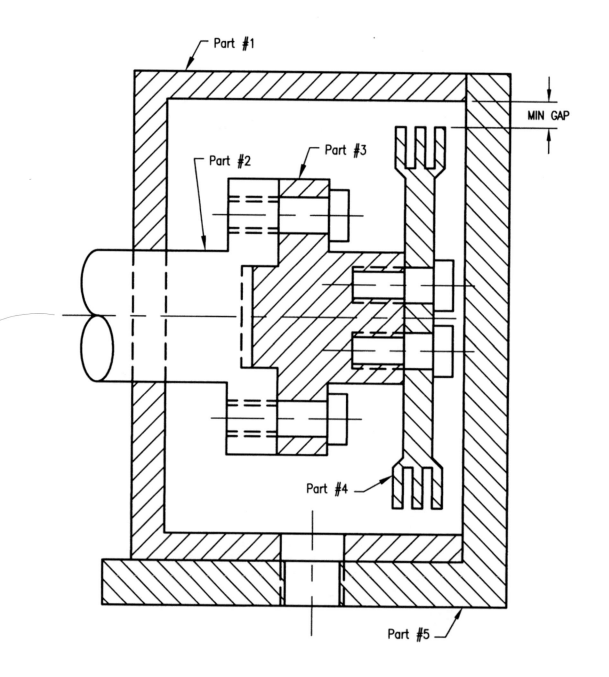

FIGURE 9-2 [Five-Part Assembly/Part #1]

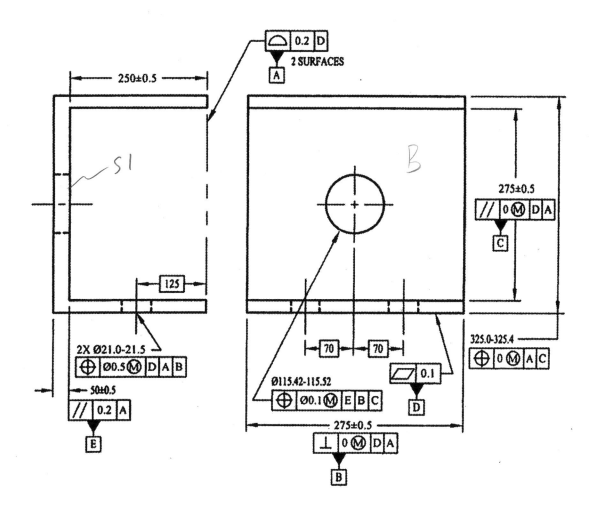

FIGURE 9-3 [Five-Part Assembly/Part #2]

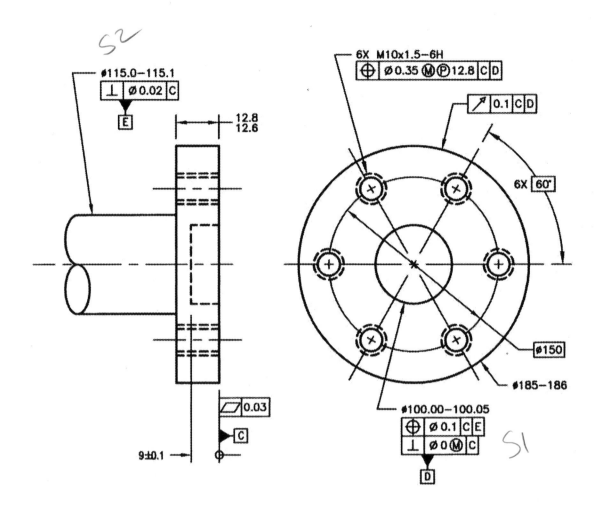

Datum feature E mates with Part #1, so it will be a factor in minimizing the gap we are calculating. The LMC of datum feature E would allow the most airspace between E and the central bore on Part #1. If E was out of perpendicularity, and E was a press fit to the bore on Part #1, we would have to consider the perpendicularity tolerance on E as a factor. It would be trigonometrically calculated as to how much it would affect the entire assembly. For example, if datum feature E was out of perpendicularity 0.02 over 100 millimeters (the length of E for this explanation only), and the assembly beyond E to the right was 1000 millimeters long (the length assumed for this explanation only) then the minimum gap would be affected by 0.2. But, since E is not a press fit to the bore, the rest of the assembly does not have to take on the angle of E, therefore the perpendicularity of E will not be considered a factor in this calculation.

The proportions, trigonometric factors and algebraic calculations of tolerance stack-up analysis will be discussed in Chapter 10. These factors, if taken to their extreme conclusions, can significantly impact the tolerance accumulation for assemblies, especially assemblies with many parts. Again, see Chapter 10 for further clarification.

FIGURE 9-4 [Five-Part Assembly/Part #3]

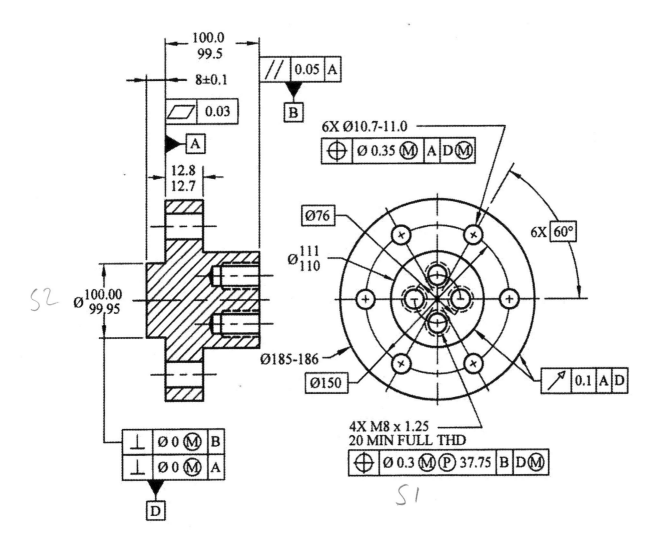

FIGURE 9-5 [Five-Part Assembly/Part #4]

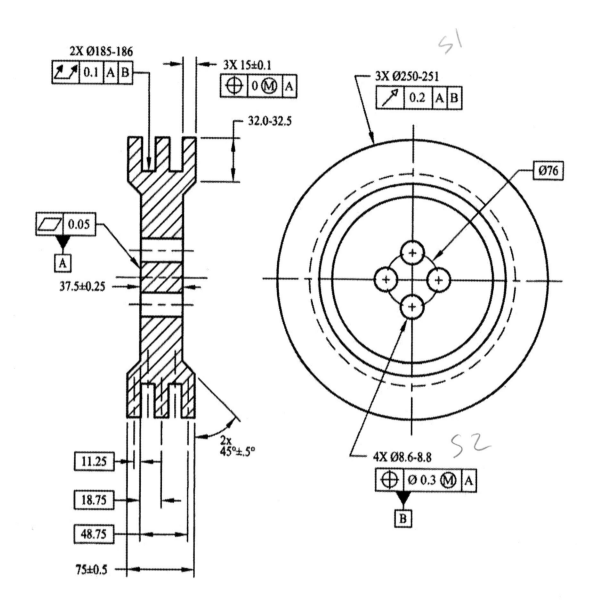

FIGURE 9-6 [Five-Part Assembly/Part #5]

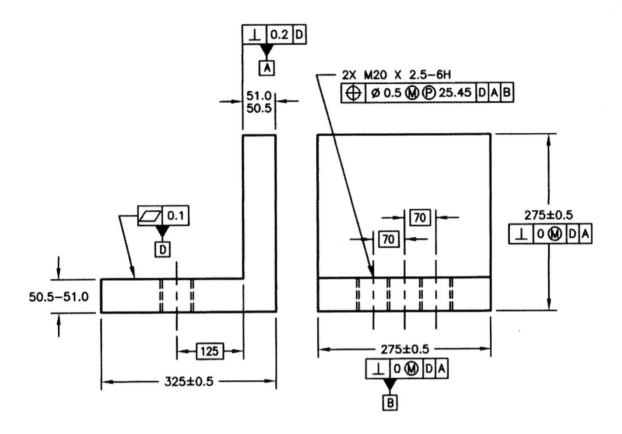

This five-part assembly creates a wide variety of analysis opportunities for each single part and for how they react together. To begin, let's look at the (side) view of the assembly.

Three of the parts are enclosed by the housing. The housing is comprised of the other two parts. To determine if the housing is large enough to contain the three parts, let's consider the cavity size. Since the holes that bind these parts together have the same basic 125mm dimension coming from datum features that seat against one another, and the virtual condition of the holes is 20 + 0.5 on the screws mounted in the threaded holes (or Ø20.5 virtual condition screws) and 21 - 0.5 on the clearance holes (or Ø20.5 virtual condition holes), we can assume this line fit is not a factor in the minimum gap for this assembly. It would be a factor in the maximum gap because of allowed airspace created by larger holes moved away from the wall (or positioned at perfect location) and smaller screws positioned away from the wall (or at perfect location).

To figure minimum gap, though, we can forget holes and work only with the sizes of the pertinent lengths. This shortest pertinent length on Part #1 is 250 – 0.5 or 249.5. Since Part #5 provides a satisfactory seating area for Part #1 (324.5 - 51.0 = 273.5) is not a factor. So, the only number important for a MIN GAP calculation from Parts #1 and #5 is the cavity minimum depth of 249.5. The parallelism tolerance on datum feature E is a factor. The profile tolerance is also a factor in that it is also a perpendicularity tolerance since it references datum D. Angular variations have the effect of closing the area available to house Parts #2, #3 and #4 of the assembly. Together the parallelism tolerance and the profile tolerance would allow the walls to lean 0.2 each, but since the size tolerance of the overall box width may not be violated, the total effect is to close the box of 0.2. See FIGURE 9-7 for an explanation.

FIGURE 9-7

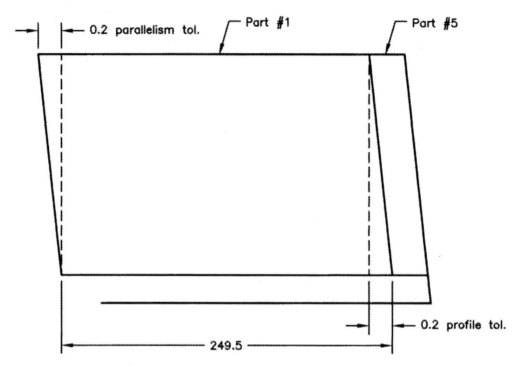

For Parts #2, 3 and 4, the stack-up analysis begins by adding the appropriate MMC sizes:

Part #2 = 12.8 MMC

Part #3 = 100.0 MMC

Part #4 is more complicated, but leads us to originate from datum plane A and go the basic dimension of 48.75 to the centerplane of the last 15±0.1 feature and adding one-half of its outer boundary. Its outer boundary is:

$$\begin{array}{r}15.1 = \text{MMC} \\ +\ 0\ = \text{Geo. Tol. at MMC} \\ \hline 15.1 = \text{Outer Boundary}\end{array} \qquad \frac{15.1}{2} = 7.55$$

So,

12.8 + 100 + 48.75 + 7.55 = 169.1

169.1 is needed to house Parts #2, #3 and #4. Since we have 249.3, we have 80.2mm of clearance. The parallelism tolerance on datum feature B on Part #3 is not a factor since it is contained within the MMC of the overall size of the pertinent features.

FIGURE 9-8

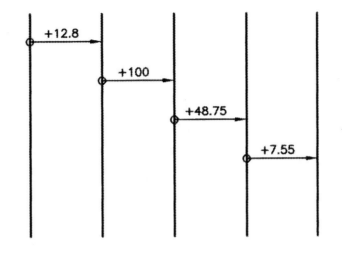

12.8 + 100 + 48.75 + 7.55 = 169.1

Since we seem to have plenty of clearance in that direction, let's examine another.

FIGURE 9-9

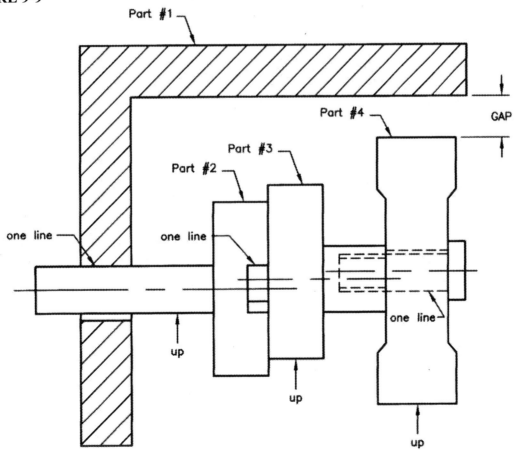

Shown with Parts #2, #3 and #4 pushed up to close the GAP with Part #1.

Part #4
 Step 1: Outside Diameter
  251.0 = MMC
  + 0.2 = Geo. Tol.
  Ø251.2 = Outer Boundary

  $\dfrac{251.2}{2}$ = R125.6 Shaft

 Step 2: Clearance Hole
  Ø8.8 = LMC

  $\dfrac{8.8}{2}$ = R4.4 Hole

Part #3
 Step 1: Threaded Hole Treated as Mounted Screw
  7.76 = LMC Screw
  - 0.30 = Geo. Tol.
  7.25 = Inner Boundary Screw
  - 0.05 = Pattern Shift due to D∂ Reference
  Ø7.41 = Inner Boundary with Pattern Shift

  $\dfrac{7.41}{2}$ = R3.705 Shaft

 Step 2: PROTRUSION
  Ø99.95 = LMC of D

  $\dfrac{99.95}{2}$ = R49.975 Shaft

Part #2
 Step 1: RECESS
  100.05 = LMC of D
  + 0.10 = Position Tol. of D
  Ø100.15 = Outer Boundary of D

  $\dfrac{100.15}{2}$ = R50.075 Hole

 Step 2: SHAFT
  Ø115 = LMC of E

  $\dfrac{115}{2}$ = R57.5 Shaft

Part #1
 Step 1: Central Hole
  115.52 = LMC Hole
  + 0.20 = Geo. Tol. at LMC
  Ø115.72 = Outer Boundary Hole

  $\dfrac{115.72}{2}$ = R57.86 Hole

Calculated Overall Dimension

Step 1:

    4.400
  - 3.705
    R0.695 = Factor #1

Step 2:

    50.075
  - 49.975
    R0.100 = Factor #2

Step 3:

    57.86
  - 57.50
    R0.36 = Factor #3

  R125.6 = Factor #4

Step 4:

      0.695
+   0.100
+   0.360
+ 125.600
  R126.755 x 2 = 253.51   So, MAX OVERALL DIMENSION is Ø253.51

or

Step 1	**Step 2**	**Step 3**		
8.80	100.15	115.72		
- 7.41	- 99.95	- 115.00	and	Ø251.2
Ø1.39	Ø0.2	Ø0.72		

Step 4

    Ø251.20
+   1.39
+   0.20
+   0.72
  Ø253.51 = MAX OVERALL DIMENSION

This must fit into Part #1's cavity (Datum Feature C):

Step 5

  274.5 = Inner Boundary of Datum Feature C (R137.25).

Step 6

```
  274.50
- 253.51        and      20.99 = 10.495 Radial Clearance Maximum
   20.99                   2
```

FIGURE 9-10

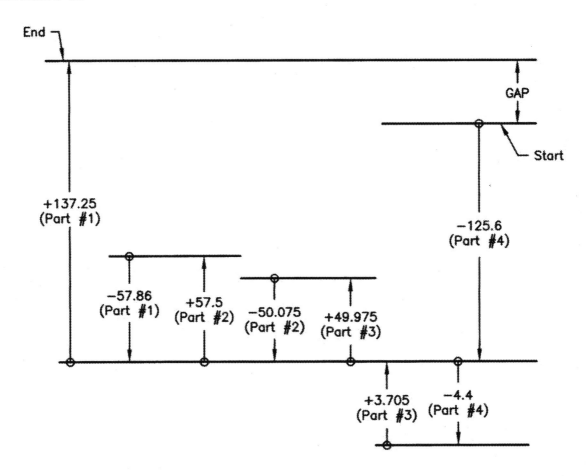

Remember...
Negative is down.
Positive is up.

−	+
125.600	
4.400	
	3.705
	49.975
50.075	
	57.500
57.860	
	137.250
−237.935	+248.430	Totals

$$+248.430$$
$$+\ -237.935$$
$$\overline{10.495} = \text{MIN. GAP}$$

The difficulty in this analysis is in deciding which features on each part are factors in the MIN GAP and which geometric controls are factors. To begin, we can look at Part #1 and determine if the inner boundary of datum feature C is a factor. The inner boundary of C is 274.5, a radius of 137.25. The outer boundary of the large clearance hole in the center of C is a factor since movement of the hole affects the parts that fit into the housing. The outer boundary is Ø115.52 + 0.20 or Ø115.72 or a radius of 57.86. The outer boundary allows the most movement of the mating shaft on Part #2.

...on Part #2

The most airspace between Part #1 and Part #2 would occur when datum feature E on Part #2 is the smallest and perfectly perpendicular to C. Therefore, the perpendicularity tolerance is not a factor in this analysis. (See the statement on the page containing Part #2 for further clarification). The LMC of datum feature E is Ø115 or a radius of 57.5. Since datum feature D is the alignment feature between Part #2 and Part #3, the threaded holes are not a factor in this analysis. Datum feature D has two geometric tolerances, but since the location of the D feature will determine the location of Part #3, the position tolerance is a factor and the perpendicularity tolerance is not. Since the largest movement (and largest size) of D allows the largest movement of Part #3, the outer boundary is calculated. The outer boundary is Ø100.05 + 0.1 or Ø100.15 or a radius of 50.075.

...on Part #3

The feature connecting Part #2 and Part #3 on Part #3 is datum feature D. Since the most airspace between them is allowed by the smallest size of D, the geometric tolerance is not a factor. The LMC of datum feature D is Ø99.95 or a radius of 49.975. Since datum feature D is the alignment feature between Part #2 and part #3, the six clearance holes on Part #3 are not a factor. The threaded holes move the screws around and create the alignment between Part #3 and Part #4. Therefore, the pattern of four threaded holes is a factor.

If the LMC screws are moved by the threaded holes as a group off of datum axis D, that would move Part #4. Also, since the pattern of threaded holes reference datum feature D at MMB, the pattern may shift an additional amount if D is produced at LMC (and we have already assumed it will be produced at LMC for this analysis). So, the inner boundary of the screws mounted in the threaded holes and shifting as a group would be: 7.76 LMC screw minus 0.3 geometric tolerance minus 0.05 pattern shift equals Ø7.41 or a radius of 3.705.

. . . on Part #4

The clearance holes connect Part #4 to the threaded holes (mounted screws) on Part #3. But since these clearance holes are positioned only to each other (and held perpendicular to datum A), and the only other pertinent features are located to the axis of this four-hole pattern, the position tolerance of these holes is not a factor. Therefore, the most movement between these clearance holes and the mounted screws from Part #3 is allowed by the largest clearance holes perfectly positioned to each other. The LMC of these holes is Ø8.8 or a radius of 4.4.

The outside diameter of Part #4 is a major factor in how Parts #2, #3 and #4 fit into the cavity on Part #1. The movement of this outer diameter off the axis of the four-hole pattern (datum pattern B) effectively increases the size necessary to fit the assembly (of Parts #2, #3 and #4) into. Therefore, the runout tolerance is a factor. The outer boundary of the outer diameter is: Ø251 + 0.2 = Ø251.2 or a radius of 125.6.

These are all of the factors in this analysis. Notice that they are all used by beginning with the factors on Part #4, exhausting them completely, then moving on to Part #3. Part #3 factors are all considered before moving on to Part #2. Part #2 factors are utilized completely before moving on to part #1 for its factors. In this way, an infinite number of parts can be used in any analysis without getting lost. It is done one part at a time. This is why doing one part analyses can prepare a person to do assembly analyses.

Still, to determine pertinent factors for any calculation, one must first decide what gap is being calculated. And second, the assembly of parts must be scrutinized to determine which features, which sizes and which geometric tolerances are and are not factors in the analysis. Again, it is worth reminding the analyst to always take the shortest route, so that unnecessary, incorrect tolerance is not accumulated for the calculation.

CHAPTER 9
EXERCISES

Exercise 9-1 [Five-Part Assembly]

PROBLEM:

Using all pertinent factors from Part #1 through Part #4, including the central hole on Part #1 as a factor, calculate the minimum clearance between Part #4's outside diameter and Part #1 of the assembly.

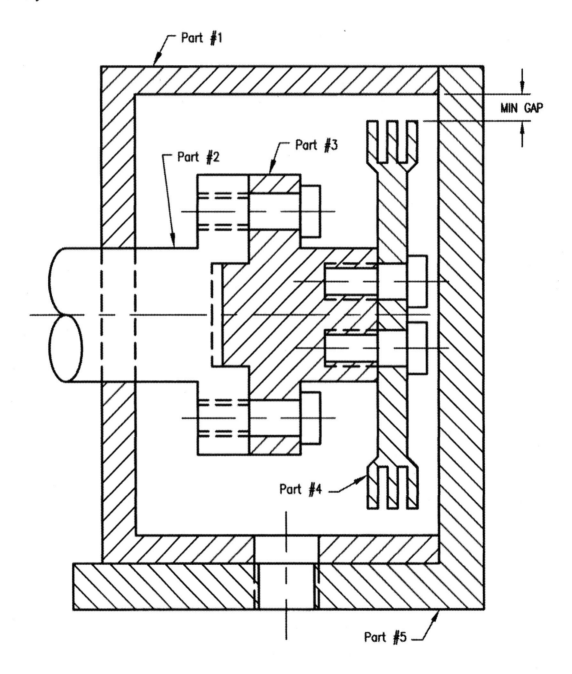

Exercise 9-1 [Five-Part Assembly, Part #1]

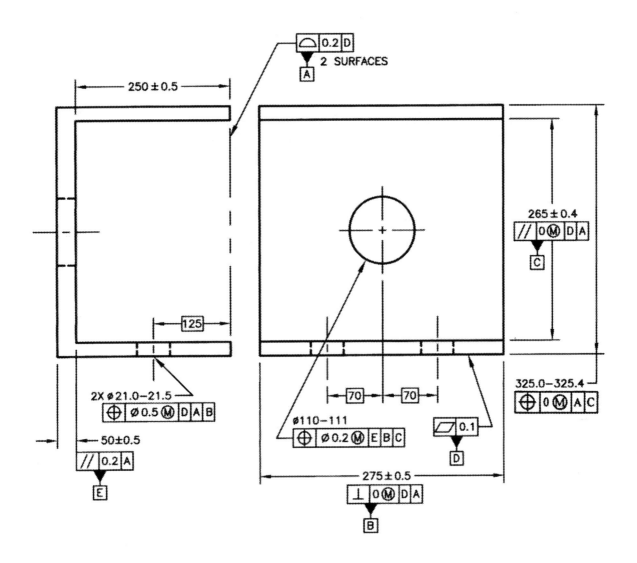

Exercise 9-1 [Five-Part Assembly, Part #2]

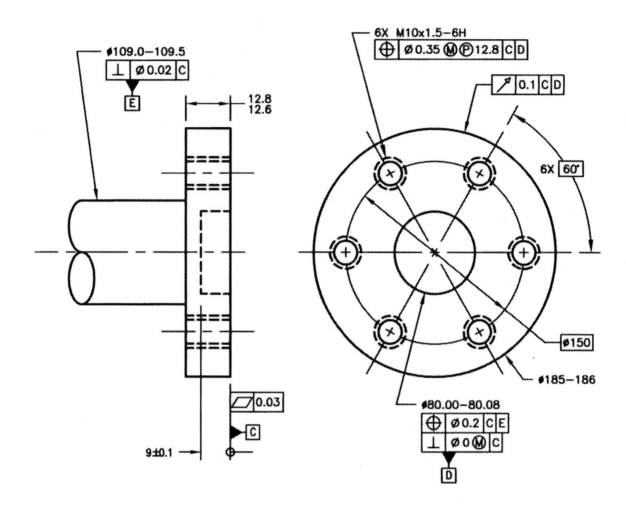

Datum feature E mates with Part #1, so it will be a factor in minimizing the gap we are calculating. The LMC of datum feature E would allow the most airspace between E and the central bore on Part #1. If E was out of perpendicularity, and E was a press fit to the bore on Part #1, we would have to consider the perpendicularity tolerance on E as a factor. It would be trigonometrically calculated as to how much it would affect the entire assembly. For example, if datum feature E was out of perpendicularity 0.02 over 100 millimeters (the length of E for this explanation only), and the assembly beyond E to the right was 1000 millimeters long (the length assumed for this explanation only) then the minimum gap would be affected by 0.2. But, since E is not a press fit to the bore, the rest of the assembly does not have to take on the angle of E, therefore the perpendicularity of E will not be considered a factor in this calculation.

The proportions, trigonometric factors and algebraic calculations of tolerance stack-up analysis will be discussed in Chapter 10. These factors, if taken to their extreme conclusions, can significantly impact the tolerance accumulation for assemblies, especially assemblies with many parts. Again, see Chapter 10 for further clarification.

Exercise 9-1 [Five-Part Assembly, Part #3]

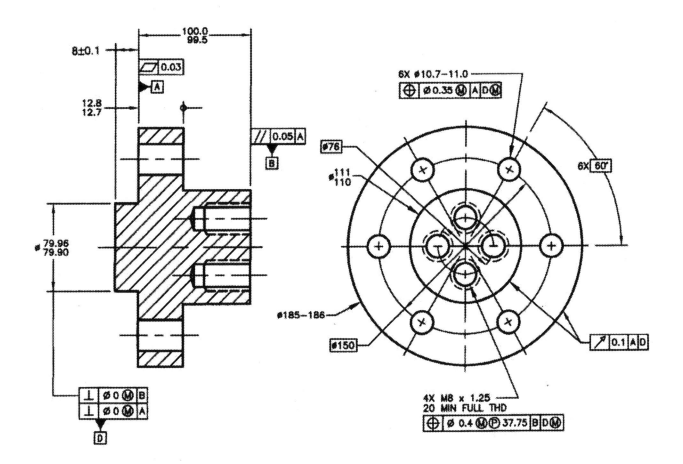

Exercise 9-1 [Five-Part Assembly, Part #4]

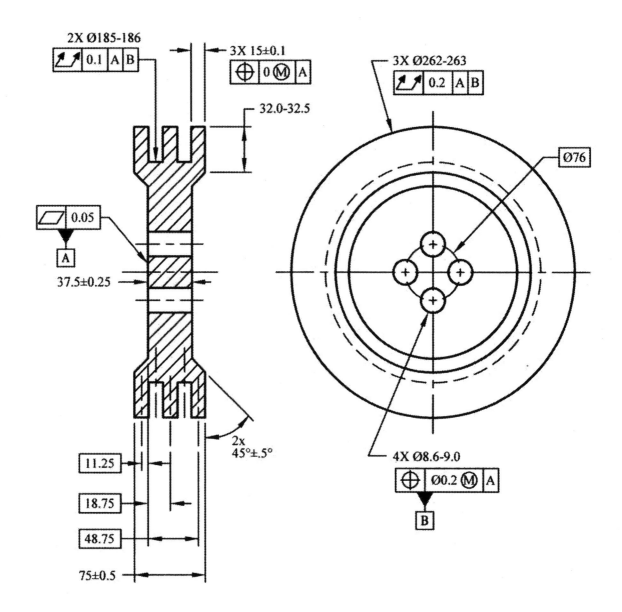

Exercise 9-1 [Five-Part Assembly, Part #5]

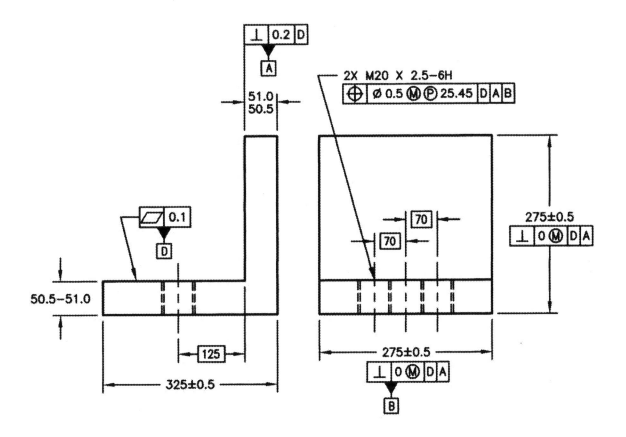

Chapter 10

TRIGONOMETRY AND PROPORTIONS IN TOLERANCE STACK-UP ANALYSIS

•Lesson Objectives:

In Chapter 10, you will:
- Learn and practice a variety of complicated stack-up problems using proportions, trigonometry and geometric tolerancing.
- Consider the effects of rocking datum features.
- Consider the rules in Y14.5.1 (*Mathematical Definition of Dimensioning and Tolerancing Principles*) for constructing a valid datum.
- Calculate the effects of differing orientations from inspection to assembly with rocking datum features.
- Consider when a computer program is needed vs. a human analyst.
- Determine how skewed vertical stacking affects horizontal housing requirements.
- Determine the affects of parts that are not stacked parallel.
- Practice formulae to calculate worst case fit conditions when trigonometry and/or proportions are a factor.
- How to blend trigonometry and algebra to determine a stack-up result.

Much is made of the trigonometric aspects of tolerance stack-up analysis. It is believed by many to be so difficult that it can only be done by computer analysis. In fact, although some of the problems should be analyzed by computer, most of them can be dealt with by simple trigonometric functions, proportions and logic. It must be remembered, though, that these aspects of the stack-up are not simple addition problems.

Chapter 10
Trigonometry and Proportions in Tolerance Stack-Up Analysis

Much is made of the trigonometric aspects of tolerance stack-up analysis. It is believed by many to be so difficult that it can only be done by computer analysis. In fact, although some of the problems should be analyzed by computer, most of them can be dealt with by simple trigonometric functions, proportions and logic. It must be remembered, though, that these aspects of the stack-up are not simple addition problems.

For example, let's consider the realistic problem of a rocking datum feature. Flatness is a geometric characteristic which can affect a stack-up analysis. In a regular geometric tolerancing approach to rocking datum features, the rocking that would be controlled by flatness would be ignored. The Y14.5 standard states that the datum feature is to be rocked to an optimum assembly condition. In other words, if it rocks, just rock it until the part checks good to simulate that the assembler would rock the part until it assembled. By that premise, the approach is that it would be illogical to rock the part until it interfered with the rest of the assembly, the same as it would be illogical to rock the part until it checked bad in inspection. But, if the rock point is in the center of the part, it is difficult (almost impossible without direction) for the inspector to determine which is the way the assembler will choose to rock it. One would have to surmise that if rocking is the option chosen (over, say shimming it up to equalize the rock) that even if there were only two ways to rock the part, there is only a 50% chance that the assembler and the inspector will choose the same way. And, in truth, there are an infinite number of ways a part may be wobbled or rocked to establish a valid primary datum plane from a datum feature that is not perfectly flat or perfectly cylindrical (or perfect in its desired configuration, whatever that is).

Still, in normal geometric tolerancing approaches, it is assumed that these chance occurrences will work out optimally. However, in tolerance stack-up analysis, the approach is exactly the opposite. If the datum feature has rock, the part is to be rocked until it interferes. So, how much the inspector and the assembler could allow the part to rock, wobble and lean (be out-of-perpendicularity, angularity or parallelism) must be calculated to determine the amount it could contribute to the possibility of interference (or whatever worst case scenario might occur). Y14.5.1 Mathematical Definition of Dimensioning and Tolerancing Principles states that a part can't rock so much that the points that establish the primary datum solely lie in one of the outer thirds of the part. But that could mean that an inspector is able to establish a valid primary datum by using only the middle third of a surface, or a little more than one of the outer thirds. It is hardly a mathematically unique solution.

FIGURE 10-1

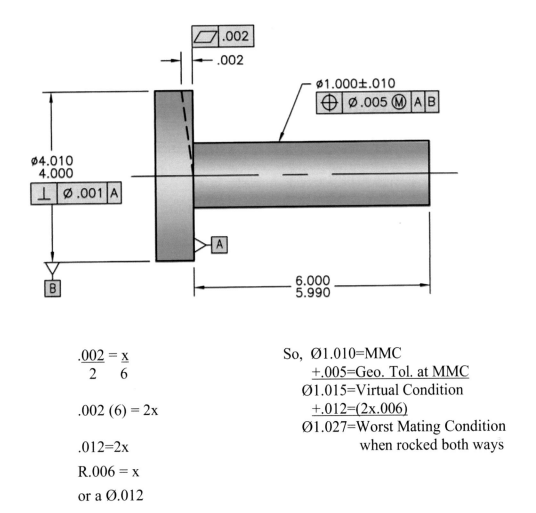

$$\frac{.002}{2} = \frac{x}{6}$$

.002 (6) = 2x

.012 = 2x

R.006 = x

or a Ø.012

So, Ø1.010 = MMC
 +.005 = Geo. Tol. at MMC
 Ø1.015 = Virtual Condition
 +.012 = (2x.006)
 Ø1.027 = Worst Mating Condition
  when rocked both ways

FIGURE 10-1 shows a simple solution using proportions. Notice that the out-of-flatness is shown all on one side of the part center. Again, this is because that is worse than it being spread out over the entire surface. In the Y14.5.1 standard, it shows that in order to be a valid primary datum feature, the points used to construct the datum plane (3 high points of contact minimum) must not lie solely in one of the outer thirds of the surface. So, although it may be possible to conceive of a slightly worse occurrence than having the rock point in the exact part center, that is the situation explained in this illustration. As the reader can see, this requires some pretending, in that, since the 1.000 inch shaft is in the center of datum feature B, a rock point could not actually be there. Still, it serves us to say that the rock point is there for this conversation. Some logic must be used to say that if the part was rocked more than that, the assembler has had to try hard to make the part not assemble and the inspector has used judgment that may not be in the best interest of the assembly in rocking the part more extremely to get it to check good.

This illustration shows a surface that has a rock point in the middle of the part. The specification allows the surface to lean by the flatness tolerance of .002. If this part is inspected on the angle of the surface that does not lean--but assembled on the surface that does lean--the 6.000 shaft will be forced to lean with it. With .002 of lean over a 2.000 radius of the flat surface, the 6.000 long shaft will be forced to lean a radius of .006.

Normally, this is ignored when calculating worst mating conditions of features like the 6.000 long shaft. It would have a worst mating condition that is most often calculated by just adding its maximum material condition and its geometric tolerance. This would make the worst case assembly condition 1.010 (MMC) plus .005 (geometric tolerance at MMC), or Ø1.015. But with the additional radial lean of .006, it would seem the worst mating condition is actually 1.015 plus 2x .006. so, Ø1.015 + .012 = Ø1.027 is actually the worst case assembly size for this shaft. Also, when calculating the minimum gap between this shaft and the housing (space) into which it fits, if we used the stack-up analysis procedure shown in this text, we would probably be working in radii. So, the calculation would take in the radial lean of the axis from center out (.006) combined with a radius of .5075 (1/2 of the Ø1.015), for a total of .5135 (1/2 of Ø1.027).

Parallelism is a factor that can be directly related to the problems that flatness creates. Parallelism, if used on a planar surface, is a control of flatness and angle. Since flatness can be thought of as applying to a surface produced with a high point in the middle, one may think of a surface as having angled down on either side of the rock point. These two halves of the surface would, therefore, not be parallel to one another and neither would they be parallel to the datum. For example, see Figure 10-2 below.

FIGURE 10-2 [Parallelism and Flatness as Factors]

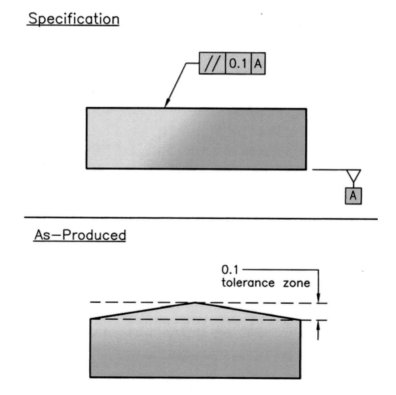

So, if two (or more) parts were stacked on top of one another, and each had the type of problem shown in the As-Produced illustration (FIGURE 10-2), the parts could exhibit problems of not fitting into assemblies because of filling too much space as an assembly or closing holes on the parts into which pins or screws had to fit. For example, see the three parts shown stacked below. The premise of this analysis is that three parts have been stacked with the left edge of each part acting as the alignment feature. This would assume that interior part features such as holes and shafts (not shown) have been positioned from the left edge as a secondary datum feature.

Each part during inspection has been adjusted (shimmed up) to allow the point shown as the high point on the bottom of parts 1 and 2 to establish the datum plane. But during assembly, the parts have been rocked over instead of equalized. This has allowed the rock point to move off of the 200 millimeter center of the parts by an insignificant amount (0.00002 on part 2 as it contacts part 1 and 0.00004 on part 3 as it contacts part 2) to represent the hypotenuse created by the out-of-flatness of the bottom of parts 1 and 2. It is simply one speculation about what could happen. Many other scenarios are also possible.

FIGURE 10-3 [Three Parts Shown Stacked]

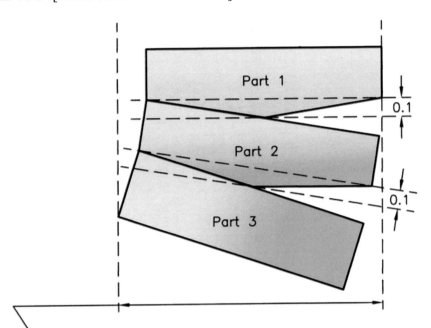

This much space is needed to house these three parts if they were stacked this way. As can be seen, the problem is compounded every time a part is added to the assembly. Also, the top of each part (shown flat here) could have their own problems and angle or rock because of out-of-flatness difficulties — which, in the case of parts 2 and 3, would compound the problem even more. Even though these problems are calculable, one can see that these situations are probably best handled by computer programs written specifically to do these stacks.

FIGURE 10-4

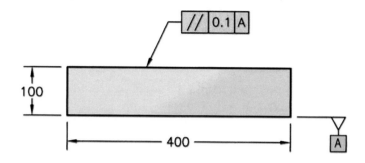

FIGURE 10-5 [Three Parts Shown Stacked with Left Edge Aligned, then the Parts are Rocked]

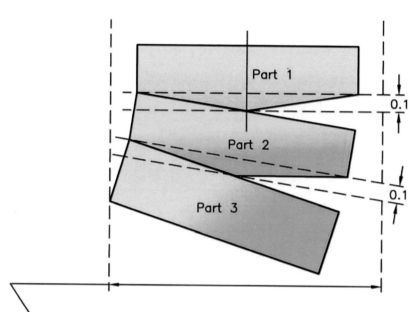

This much space would be needed to house these three parts if they were stacked with their edges aligned, then rocked in either direction. As can be seen, the problem is compounded every time a part is added to the assembly. Also, the top of each part (shown flat here) could have their own problems and angle or rock because of out-of-flatness difficulties — which, in the case of parts 2 and 3, would compound the problem even more. Even though these problems are calculable, one can see that these situations are probably best handled by computer programs written to specifically do these stack-ups. Unlike Figure 10-3, Figure 10-5 recognizes and illustrates the possibility of the parts rocking to either the left or right. To accomodate this, space has been added to both sides of the assembly.

To calculate the overall housing requirements for the three parts, we must first visualize the parts as shown in FIGURE 10-5 and calculate the offset of the left-most point on Part #3 (shown as the left-most point in Triangle 5 in FIGURE 10-6) from the left side of Part #1 (shown in FIGURE 10-5). Since the parts may rock either left or right, the offset calculated must be doubled, then added to the 400mm dimension of Part #1. Since there are no tolerances shown on the size of the width or height of the parts, the housing requirements would be the same whether the left edges, centers or right edges of the parts were aligned and then rocked. The trigonometry and algebra shown in the following pages calculate the offset and housing requirements. The logic for these calculations is then summarized.

FIGURE 10-6

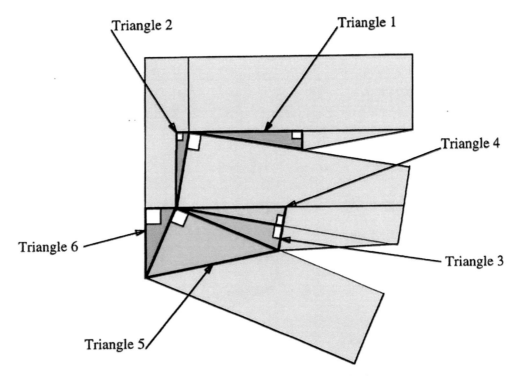

Triangles 1, 3 & 4 are identical.

Triangle 1 [FIGURE 10-7]

$a_1 = .1$

$b_1 = 200$

$c_1 = 200.000025$

$TAN\ A_1 = \dfrac{a}{b}$

$TAN\ A_1 = \dfrac{.1}{200}$

$TAN\ A_1 = .0005$

$A_1 = .028648°$

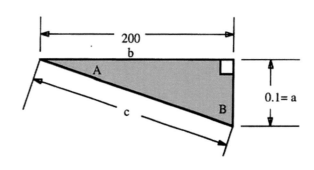

Triangle 2 [FIGURE 10-8]

$B_2 = 180° - 90° - A_1$

$B_2 = 90° - A_1$

$B_2 = 90° - .028648°$

$B_2 = 89.971352°$

$c_2 = 99.9$

$a_2 = c_2 \cdot \cos B_2$

$a_2 = 99.9 \,(\cos 89.971352)$

$a_2 = 99.9 \,(.0005)$

$a = .04995$

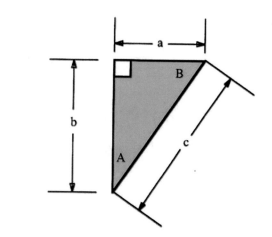

Triangle 3 [FIGURE 10-9]

$a_3 = .1$

$b_3 = 200$

$A_3 = .028648°$

$B_3 = 89.971352°$

$C_3 = 90°$

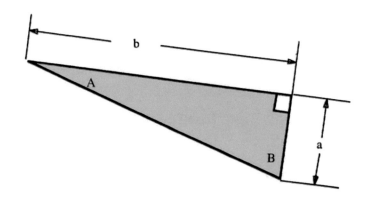

Triangle 4 [FIGURE 10-10]

$a_4 = .1$

$b_4 = 200$

$c_4 = 200.000025$

$\tan A_4 = \dfrac{a}{b}$

$A_4 = .028648°$

$B_4 = 89.971352°$

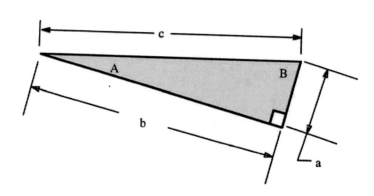

Triangle 5 [FIGURE 10-11]

$a_5 = 100$

$b_5 = 200.000025$

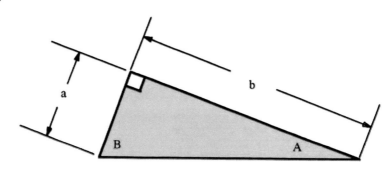

Triangle 6 [FIGURE 10-12]

$a_6 = c_6 \cdot \cos B_6$

$a_6 = 100(\cos B_6)$

$a_6 = 100(\cos 89.942704)$

$a_6 = 100(.001)$

$a_6 = .100$

$c_6 = 100$

$B_6 = 180° - 90° - 2(A_4)$

$B_6 = 90° - 2(.028648°)$

$B_6 = 90° - .057296°$

$B_6 = 89.942704$

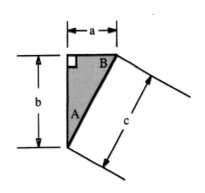

Therefore, offset equals:

$a_2 + a_6 =$ offset   .04995 + .100 = .14995 (rounded off to .150)

Housing must increase by .150 on each side:
.150 x 2 = .300   So, 400.00 + .300 = 400.300

FIGURE 10-13 [A Simplified Summary of this Problem]

So, the logic is that 3 parts are longer (taller) than 2 parts vertically. The angle is experienced by the middle part cocking its center and left edge to the left by .050 (rounded off).

The bottom part adds its height putting its center and left edge over to the left by another .050 (since it is as tall as the middle part). Then, since the middle part's bottom angles by the same amount as the top part's bottom, the bottom part has its middle and left edge cocked out to the left by an additional .050.

So, in this case, whether the centers of each part are aligned or the left edges are aligned, the result is the same. When the parts are rocked, the left edge of the bottom part moves left by .150.

The fact that no tolerances are included on the 400mm part widths simplifies the results shown here.

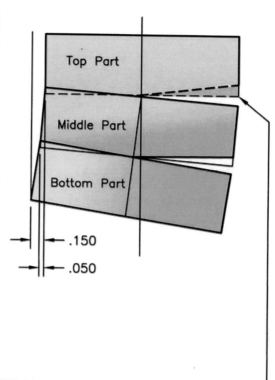

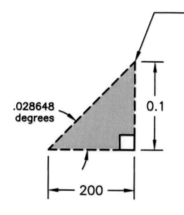

The length of the center and the edge drift is 100 times the sine of .028648.

(100 represents the height of each part).

So; 100(sin .028648 degrees)

=100(.0005)

=0.05

If the drift is 0.05 on the middle part, then on the bottom part it is 3 times 0.05, because of its angle (given by the middle part and its own height). So, it is 0.15.

So, the overall housing needs are a radius of 200 plus .15=200.15 or a diameter of 400.3.

Remember, the purpose of this exercise has been to show how complex the calculations can be when one assumes certain flatness, parallelism and perpendicularity problems may occur. It is to encourage one to write or seek out computer programs to help in the solutions to these problems. But it is also to keep in mind that computer programs may not be written to correctly consider these problems. As a good consumer, you must study the software and what the algorithms are written to consider for you to be certain they are sufficient to solve the possibilities that may occur ... and to solve for those problems that concern you the most.

The situations given in this section are practical but may not (as mentioned) represent the absolute worst part conditions. For example, ASME Y14.5.1 Mathematical Definition of Dimensioning and Tolerancing Principles basically states that when a planar primary datum feature rocks, it may not be rocked so much as to allow the three high points of contact used to construct the primary datum plane to lie solely within only one of the outer thirds of the surface. This means that a valid primary datum plane may have all three points from the surface lying in just a little more than one-third of the surface. In the examples given in this section, one-half of the surface was used and, although it is practical to assume one would not want to use less than one-half, it is legal to do so. This means, of course, that the situation given in this section could have been worse - and the math could have been harder.

So, please take this section as it was intended. It was to show that the angular difficulties created by surface problems, similar to the ones used here, can make the mathematics so difficult that one may not wish to try to solve them without the help of a computer program.

If surfaces were only out-of-parallel and not out-of-flat, one can calculate how much that would affect an assembly for certain considerations. Let's assume we had a three-part assembly and all parts were cylindrical. If we had a shaft that had to fit into a central hole in all three parts, the parallelism of the top and bottom of the part sandwiched between the other two would be a factor. For example, see FIGURE 10-14.

FIGURE 10-14 [Parallelism Only as a Factor]

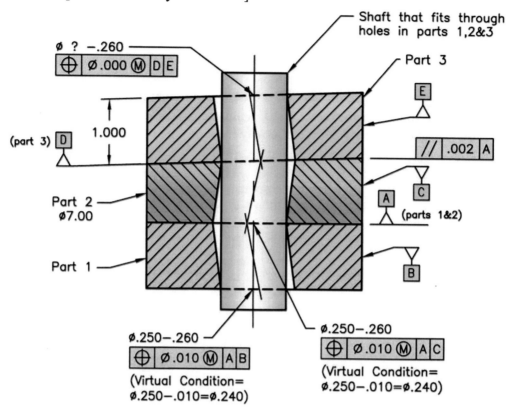

The virtual condition of the hole in Part #3 would have to increase beyond the Ø.240 virtual condition boundary shared by Parts #1 and #2 because of the out-of-parallelism between the top and bottom of Part #2. The amount of increase in terms of a diameter can be calculated by the following formula:

$$\frac{\text{Parallelism Tolerance between Datum D feature and Datum A}}{\text{Diameter of Part \#2}} = \frac{\text{Increase in Part \#3's Hole Virtual Condition}}{\text{Thickness of Part \#3}}$$

So, in this problem:

$$\frac{.002}{7} = \frac{x}{1}$$

x = .0002857 = Increase in the Virtual Condition of Part #3's Hole

So, the ? (unknown size) listed above as the MMC and Virtual Condition of the hole in Part #3 is Ø.2400 + .0003 = Ø.2403.

As is sometimes the case, the increase is so small, it may be that once calculated, it is ignored. One should be aware however, that the more parts are stacked on top of one another, the more significant the increase becomes and at some point can no longer be ignored.

Trigonometric Functions

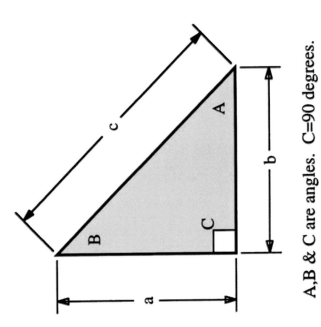

A, B & C are angles. C = 90 degrees.

Formulas	Sides & Angles Known							
	Side a; Side b	Side a; Hyp. c	Side b; Hyp. c	Hyp. c; Angle B	Hyp. c; Angle A	Side b; Angle B	Side b; Angle A	Side a; Angle B	Side a; Angle A
	$c=\sqrt{a^2+b^2}$	$b=\sqrt{c^2-a^2}$	$a=\sqrt{c^2-b^2}$	$b=c\cdot\sin B$	$b=c\cdot\cos A$	$c=b/\sin B$	$c=b/\cos A$	$c=a/\cos B$	$c=a/\sin A$
	$\tan A=a/b$	$\sin A=a/c$	$\sin B=b/c$	$a=c\cdot\cos B$	$a=c\cdot\sin A$	$a=b\cdot\cot B$	$a=b\cdot\tan A$	$b=a\cdot\tan B$	$b=a\cdot\cot A$
	$B=90°-A$	$B=90°-A$	$A=90°-B$	$A=90°-B$	$B=90°-A$	$A=90°-B$	$B=90°-A$	$A=90°-B$	$B=90°-A$

CHAPTER 10
EXERCISES

Exercise 10-1

Calculate the <u>increase</u> and the <u>virtual condition</u> for the hole in Part #3, given the following changes to FIGURE 10-14.
    a) the parallelism tolerance on Part #2 to datum A = .020
    b) the diameter of Part #2 = 36.000
    c) the thickness of Part #3 = 16.000

STATISTICAL TOLERANCING

• Lesson Objectives:

In Chapter 11, you will learn to:
- Convert an arithmetically calculated tolerance to a statistically calculated tolerance.
- Use the Root Sum Square formula.
- Determine the amount of tolerance assigned in an assembly statistically likely to be consumed.
- Compare the amount of tolerance statistically likely to be consumed to the arithmetically calculated tolerance.
- Calculate the percentage that each tolerance may be increased by to allow the assembly to statistically consume only the arithmetically assigned assembly tolerance.
- Use a 'Safety' or 'Correction' numerical factor as a multiplier in the Root Sum Square formula.
- Reintegrate the Statistical Tolerance into the assembly.

Chapter 11
Statistical Tolerancing

The probability of producing features at their worst-case assembly conditions is unlikely unless the manufacturers are targeting them. In most cases, they are not. And, although manufacturing practices differ from place to place, even if they were shooting for the smallest holes and the largest shafts, they would unlikely be trying to use up all of the tolerances that affect assembly.

In part geometry there are, at most, four things that would have to come together to create the worst-case assembly conditions. These are size, shape (3D form), angle (orientation), and location. For example, in mating features that have size dimensions and tolerances and that use position tolerances, all four would affect their worst case. Under ANSI standards and the ASME Y14.5-2009 Dimensioning and Tolerancing Standards, size tolerances also control form tolerances on all rigid features. Since position controls both angle and location, a feature controlled for size, and consequently form, that is also positioned has to span all of its tolerances of size, shape, angle and location to be made at the worst-case assembly condition.

Manufacturers that use methods of Statistical Process Control well are generally believed to be in statistical control instead of the statistical chaos that it would take to produce features at their worst. They will instead produce parts which, when measured, will be found to follow a natural variation which form a natural bell curve distribution of part dimensions. They will commonly show that a large percentage of the produced parts measure close to the average dimension. The magnitude and spread of all dimensions will vary from the average by an amount that can be represented in a graph known as a Gaussian frequency curve. This curve is commonly referred to as a normal bell-shaped curve, wherein the area under the curve represents 100% of the parts produced. The height of the curve represents the times dimensions have been produced for the variable individual components. The dispersion of the dimensions under the curve is commonly described with the term "Standard Deviation", often represented by the Greek letter for sigma (σ).

The relationship between the standard deviation and the area displayed under the curve can be shown as a percentage of the produced parts. The arithmetic mean plus or minus one standard deviation is often described as containing 68.26% of the produced parts in a Gaussian frequency curve. The arithmetic mean plus or minus two standard deviations is 95.46% of the total production. The arithmetic mean plus or minus three standard deviations is 99.73%. The limit of plus or minus three sigma is defined as the 'natural tolerance'. See FIGURE 11-1.

FIGURE 11-1 [Standard Bell-Shaped Curve]

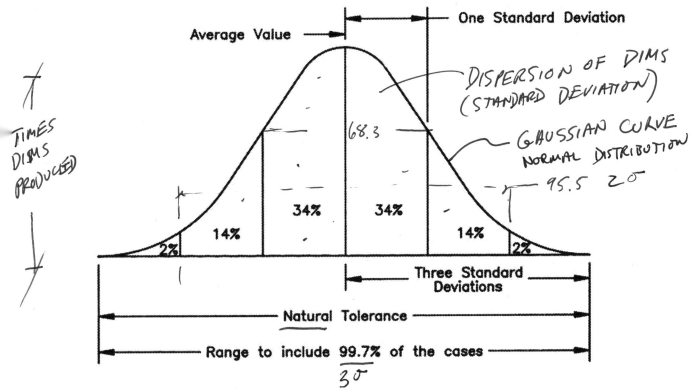

Statistical probability can be applied to tolerance stack-up analysis for assemblies both with and without geometric tolerances. To establish tolerances using this method, we can use a formula based on the Pythagorean Theorem. It can be expressed as 'the tolerance of the assembly is equal to the square root of the sum of the squares of the individual component tolerances'. This formula is called the 'root sum square' or RSS formula. See FIGURE 11-2 below.

FIGURE 11-2

$$T_A = \sqrt{T_1^2 + T_2^2 + T_3^2 + \ldots T_N^2}$$

Statistical probability mathematics have been long practiced and well documented. They are most reliable when a large production run of component parts are being manufactured. When small production runs are being used, the frequency curves tend to be skewed from what has been described here as normal.

Once the calculations described in this book for 100% tolerance stack-up analysis have been performed, the following steps can be incorporated to gain tolerance for the assembly and its components:
  a) Using the derivation of the Pythagorean Theorem, called the RSS (root sum square) formula, calculate the statistical probability for the assembly.
  b) Determine the percentage ratio between the statistical probability tolerance and the previously calculated 100% assembly tolerance

c) Determine the increased statistical probability tolerances to be redistributed to the assembly's component parts.

For examples, we will use a few pages from this textbook. See FIGURE 11-3, FIGURE 11-4 and FIGURE 11-5. From Chapter 2, 10 plates fit into a box. MIN and MAX gaps are calculated.

FIGURE 11-3   [Box Assembly]

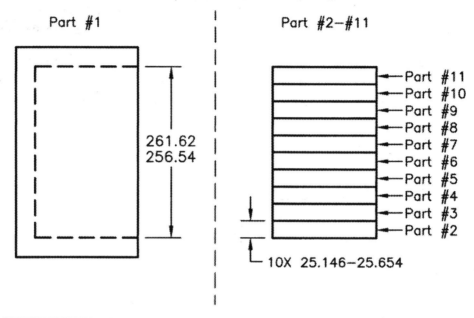

FIGURE 11-4 [Loop Analysis and Numbers Diagram]

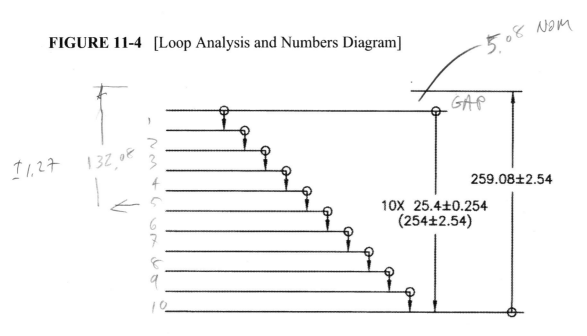

FIGURE 11-5 (Each Tolerance is Squared and the Sum of the Squares is Found)

	−	+	± Tol.	± Tol. squared	
1	25.4		0.254	0.064516	Part #11
2	25.4		0.254	0.064516	Part #10
3	25.4		0.254	0.064516	Part #9
4	25.4		0.254	0.064516	Part #8
5	25.4		0.254	0.064516	Part #7
6	25.4		0.254	0.064516	Part #6
7	25.4		0.254	0.064516	Part #5
8	25.4		0.254	0.064516	Part #4
9	25.4		0.254	0.064516	Part #3
10	25.4		0.254	0.064516	Part #2
		259.08	2.540	6.451600	Part #1
	254.00	259.08	5.08	7.096760	Totals

259.08
− 254.00
+5.08 = difference between (technically the sum of) the positive and negative mean dimensions

7.096760 = Sum of the Squares (of the tolerances)

$\sqrt{7.09676}$ = **2.6639744 = Square Root of the Sum of the Squares (Root Sum Square-RSS)**

$\dfrac{2.6639744 \text{ (tol. likely to be consumed)}}{5.08 \text{ (arithmetically calculated tol.)}}$ = 0.5244044   So, 2.6639744 is about 52% of 5.08.

$\dfrac{1}{0.5244044}$ = 1.9069252 and 1.9069252 × 2.6639744 = 5.08 (the originally calculated arithmetic tolerance)

So, the arithmetically calculated 100% tolerance allows a gap that is 5.08±5.08. This is a maximum gap of 5.08 + 5.08 = 10.16. And it allows a minimum gap that is 5.08 − 5.08 = 0.

The statistically calculated assembly tolerance predicts a likely gap that is 5.08±2.66 (rounded off to two decimal places). The ±2.66 is the amount of tolerance likely to be consumed in a natural bell curved distribution of manufactured parts for this assembly. So, if we want to consume ±5.08 tolerance, the piece part tolerance should be increased to 191% (rounded off from 190.69252%).

So, the ±0.254 (the tolerance given originally to each of Part #2 through Part #11) becomes 1.91 x 0.254 = 0.48514 or a tolerance for Part #2 through Part #11 of ±0.485 (rounded off to three decimal places). Likewise, the tolerance for Part #1 becomes 1.91 x 2.54 = 4.8514 or a tolerance of ±4.85 (rounded off to two decimal places). This is the answer to the problem of what the statistical tolerances for each part in the assembly would be if calculated by the RSS method.

Since statistically the tolerance originally assigned would not be fully consumed, arithmetically the gap calculation given, 5.08±5.08, becomes a statistical probability within ±3 sigma of consuming only 5.08±2.66. This leaves a maximum gap likely of 5.08 + 2.66 = 7.74. And it leaves a minimum gap likely of 5.08 - 2.66 = 2.42. **This prediction applies before the tolerances are increased to the statistical tolerances.**

So you can reduce GAP (NOM) AND GET 3σ

Given the newly assigned statistically calculated tolerances, with each piece part given a tolerance of ±0.485 for Part #2 through Part #11 for a total of ±4.85 and Part #1 with a tolerance of ±4.85, we have the mathematical possibility of a 4.85 + 4.85 tolerance of ±9.7. So, we could have a gap of 5.08±9.7 for a maximum gap of 14.78 and a minimum gap of -4.62. In other words, an interference of a maximum of 4.62 is possible, but unlikely.

More likely is that only ±5.08 of the ±9.7 tolerance will be consumed by manufacturing. Part #2 through Part #11 will be assigned ±0.485 tolerance.

GAP NOM + TOL

$0.485^2 = 0.235225$

$0.235225 \times 10 = 2.35225$

Part #1 will be assigned ±4.85 tolerance.

$4.85^2 = 23.5225$.

And, $2.35225 + 23.5225 = 25.87475$ = Sum of the Squares of the Statistical Tolerances

The square root of 25.87475 = **5.08=RSS of the Statistical Tolerances**

So, by the same RSS (root sum square) method that we arrived at the ±9.7 tolerance, we were able to calculate that the likely consumed amount of tolerance by the assembly will be only ±5.08. So, the likely maximum gap is still 10.16 and the likely minimum gap is still zero, even though the tolerance has been increased to 191% from ±5.08 to ±9.7.

For another example of this methodology, we will use an illustration from Chapter 6, where in a fixed fastener two-part assembly we calculated the MIN GAP for the lower left. To do this, we had to choose the correct route to follow, convert the dimensions to equal bilateral with plus or minus tolerances, then calculate the gap. We used basic dimensions that had tolerances expressed as zero and slots and tabs whose tolerances included both size and position. Still, in the end, we came up with a series of plus and minus tolerances that we used to calculate the minimum gap. See the following illustrations.

```
     .322
   +1.270
   ──────
    1.542

5 × .0645 = .322
1 × (.322)² = .104
```

OR REDUCE GAP

±.508 TO ±2.66 ⟹ REDUCE 259.08 BY 2.42

```
  259.08
 -  2.42
 ───────
  256.66
 -254.00
 ───────
    2.66  MAX STAT GAP
```

FIGURE 11-6 [Assembly Drawing from Chapter 6]

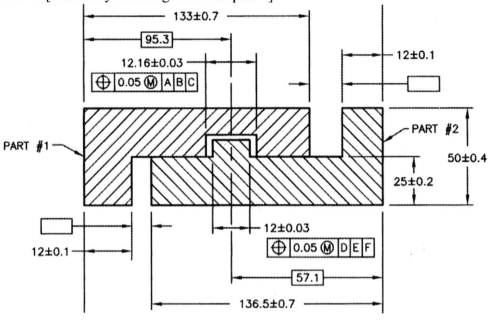

FIGURE 11-7 [Calculated Minimum Gap Lower Left]

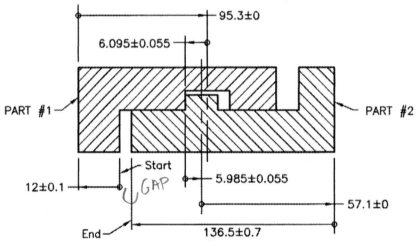

Right to Left	Left to Right	±Tol	
−	+		
12.000		0.100	Wall
	95.300	0.000	Basic Dim
6.095		0.055	Slot
	5.985	0.055	Tab
	57.100	0.000	Basic Dim
136.500		0.700	Overall Dim
154.595	158.385	0.910	Totals
158.385 −154.595 3.790	3.79 −0.91 2.88		MIN GAP

3.79
−.71
3.08 MORE PROBABLE

2.88 +.20 = 3.08 MIN GAP

Now we will use that illustration to show how we can convert to statistical tolerances with the same methods we used on the previous example that had only plus or minus tolerances.

FIGURE 11-8

−	+	±	± squared	Type
12.000		0.100	0.01	Wall	(1)
	95.300	0.000	0.00	Basic Dim	(2)
6.095		0.055	0.003025	Slot	(3)
	5.985	0.055	0.003025	Tab	(4)
	57.100	0.000	0.00	Basic Dim	(5)
136.500		0.700	0.49	Overall Dim| (6)
154.595	158.385	0.910	0.50605	Totals

    158.385
  − 154.595
   +3.790 = difference between (technically the sum of) the positive and negative   NOM GAP
           mean dimensions

0.50605 = Sum of Squares

$\sqrt{0.50605}$ = **0.7113719** = Square Root of the Sum of the Squares (RSS)

$\dfrac{0.7113719 \text{ (tol. likely to be consumed)}}{0.91 \text{ (arithmetically calculated tol.)}} = 0.7817273$ → 78.2% OF WC TOL

So, 0.7113719 is about 78% of 0.91.

$\dfrac{1}{0.7817273} = 1.2792184$ and $1.2792184 \times 0.7113719 = 0.91$ (the originally calculated arithmetic tolerance)

The arithmetically calculated 100% tolerance allows a minimum gap that is 3.79 − 0.91 = 2.88

NOM − WCT = 2.88

.91 − .71 = .20 ⇒ CAN REDUCE GAP BY INCREASING DIM OF FEATURE

(WC)

The statistically calculated assembly tolerance predicts a likely minimum gap that is 3.79 - 0.71 (rounded off to two decimal places). This 0.71 tolerance is the amount of tolerance likely to be consumed in a natural bell curved distribution of manufactured parts for this assembly. So, if we want to consume ±0.91 tolerance, the piece part tolerances should be increased to 128% (rounded off from 127.92184%).

So, the ±0.055 tolerance for the Slot and the Tab becomes 1.28 x 0.055 = 0.0704 or a tolerance for each that is ±0.070 (rounded off to three decimal places). The tolerance for the Wall becomes 1.28 x 0.100 = ±0.128. The tolerance for the Overall dimension becomes:
1.28 x 0.7 = ±0.896. <u>This is the answer to the problem.</u>

Since statistically the tolerances originally assigned would not be fully consumed, the minimum gap calculation given originally as 3.79 - 0.91 = 2.88 MIN GAP becomes a statistical probability within ±3 sigma of consuming only 3.79 - 0.71 = 3.08 MIN GAP.

Given the newly assigned statistically calculated tolerances, with each piece part given a statistically calculated tolerance, we have a mathematical possibility of a MIN GAP that is 3.79 minus the sum of the statistical tolerances. They are: 0.070 (Slot) + 0.070 (Tab) + 0.128 (Wall) + 0.896 (Overall dim.) = ±1.164.

    0.070 (slot)
  + 0.070 (Tab)
  + 0.128 (Wall)
  + 0.895 (Overall dim.)
    ±1.164

$\frac{.91}{.71} = 1.28$

$.91 \times 1.28 = 1.164$

And it leaves a MIN GAP of 3.790 - 1.164 = 2.626. So, arithmetically we might have a MIN GAP that is 2.626, but this is highly unlikely.
• The Slot and Tab statistical tolerances of 0.07 when squared become 0.0049 each.
• The Wall statistical tolerance of 0.128 when squared becomes 0.016384.
• The Overall dimension statistical tolerance of 0.898 when squared becomes 0.806404.
• When added, these tolerances are:
    0.0049
  + 0.0049
  + 0.016384
  + 0.806404
    0.832588=The Sum of the Squares of the Tolerances

• The square root of 0.832588 = 0.91=Square Root of the Sum of Squares (RSS rounded off).

So, we have again shown by the RSS (Root Sum Square) formula that even though the statistical assembly tolerance (±1.164) is greater (increased to 128%) than originally calculated arithmetically as ±0.91, the amount of tolerance likely to be consumed is still only ±0.91.

This statistical approach assumes a <u>zero mean shift</u> for all the dimensions being used. It is based on manufacturing processes that are <u>in statistical</u> control, not in statistical chaos. Those not employing Statistical Process Control in the manufacture of the workpieces should not use the RSS tolerancing methodology described here, since the results would not be valid.

Also, the RSS model assumes that parts produced for the assembly have been mixed and the components picked at random. The logic of the RSS Model is interesting. It basically allows more tolerances for those manufacturers that need it least--those using SPC controls. It calculates that the chances of producing a part that spans its larger Statistical Tolerance (ST) are so small that if it does happen, the randomly selected mating parts will make up for the potential problem by not spanning their tolerances. And, in fact, it presupposes that the mating part will be produced so much better than its tolerance extremes as to allow the parts to assemble well. If this is a false assumption, unacceptable and perhaps non-functional conditions may arise, such as interference of material. In general, the RSS method is not used if there are less than four dimensions in the stack-up analysis.

The 100% tolerancing method scares many professionals when they see that a line-fit possibility exists between mating features. This happens when the inner boundaries of holes (or slots) and the outer boundaries of shafts (or tabs) that are mating features are the same. Professionals calculating the worst mating conditions of such features can see the line-fit possibility and sometimes are uncomfortable with this. If that makes them uncomfortable, then allowing more tolerance using the RSS calculations, and consequently a greater possibility of interference, should make them even more unsettled.

Back in 1968, a man named A. Bender wrote a paper for SAE (Society of Automotive Engineers) entitled: *Statistical Tolerancing as it Relates to Quality Control and the Designer*. In this paper, he suggested a safety factor be added to the RSS formula. Instead of just taking the square root of the sum of the squares of the individual feature tolerances, he suggested a factor of 1.5 be multiplied by the answer of the standard RSS solution. In other words, 1.5 times the square root of the sum of the squares of the individual feature tolerances. This was so that the additional tolerance given to the piece parts in the assembly was not quite so risky. This formula predicts a larger portion of the arithmetically calculated tolerance is likely to be consumed by manufacturing. So, if the tolerance of the features is increased to be a Statistical Tolerance, it isn't increased as much as it might have been using the RSS formula without a "safety factor" of 1.5. It is known that, in most cases, producing features at their worst-case condition is unlikely, but it is also known that it happens. Some studies have shown that the RSS methodology doesn't accurately reflect the reality of what is produced, so to have a 'cushion' would be wise.

Root Sum Square Formula with the Bender Safety Factor

$$T_A = 1.5\sqrt{T_1^2 + T_2^2 + T_3^2 + \ldots T_N^2}$$

In the years since 1968, many authors and statisticians have suggested other 'safety' or 'correction' factors. They are often based on studies they have done that include probable or measured repeatability and/or accuracy rates of particular manufacturing processes used on specific products. These 'correction' factors often range from 1.4 to 1.8, although one of the most common remains the 1.5 suggested by Bender all those years ago. Before trying an arbitrary 'correction' factor, it is wise to look to your company to see if they have established which, if any, of these factors have been approved.

To practice any of the methods shown in this unit, pick examples from earlier units in this book and calculate the statistical tolerances with the RSS method with or without 'correctional' factors such as 1.5.

Reintegrating the Statistical Tolerance into the Assembly

Once the Statistical Tolerance has been calculated, it has to be integrated back into the assembly. In the case of the first example with only plus or minus tolerances, it is a simple process. The mean dimensions remain the same and instead of the tolerances being ±0.254 on Parts #2-#11, they will be ±0.485. On Part #1 the tolerance is just changed from ±2.54 to ±4.85.

With the second example, the process of integrating back in the Statistical Tolerance is a little more difficult. Since the boundaries for the Slot and Tab on which the tolerances are based include both size and geometric tolerances, we have to include both in our reintegration. The Wall and the Overall dimensions are just size, so they will be easy. Remember, the tolerances were increased to approximately 128%. If we keep the same mean dimensions for the Wall, it will just be 12±0.128, instead of 12±0.1. For the Overall dimensions, it will be 136.5±0.896, instead of 136±0.7.

The Slot will be 6.095±0.07, instead of 6.095±0.055, and the Tab will be 5.985±0.07, instead of 5.985±0.055. But with both of these, the tolerances have to be distributed between the size and geometric position tolerance. To do this, we can try to reverse the process with which we began the problem. We first established inner and outer boundaries, determined a mean dimension and divided the difference by two to get the equal bilateral tolerance. If we keep the same mean, we can determine the new inner and outer boundaries by adding and subtracting the Statistical Tolerances.

So, in the case of the Slot, we can begin by saying 6.095 x 2 = 12.19. And then multiply the Statistical Tolerance by 2 to get 0.07 x 2 = 0.14. So, the dimension and tolerance become 12.19±0.14. The inner boundary becomes 12.19 - 0.14 = 12.05. The outer boundary becomes 12.19 + 0.14 = 12.33. Since the original geometric tolerance was 0.05 at MMC, we can increase this to 128%, which would make it 0.064. The geometric tolerance at LMC was 0.11, which increased to 128% is 0.1408.

Now, if we add 0.064 to the inner boundary of 12.05 we get the new MMC of the Slot which is 12.114. And if we subtract 0.1408 from the outer boundary of the Slot, which is 12.33, we get the new LMC which is 12.189 (rounded off). So, the new specification of the Slot size is 12.114 - 12.189. And its new position tolerance is 0.064 at MMC.

We can determine that this calculation is correct by calculating the new inner and outer boundaries given these new specifications. The inner boundary is 12.114 (MMC) - 0.064 (geo. tol. at MMC) = 12.05 inner boundary. This is just as we determined it should be. The outer boundary is 12.189 + 0.064 (geo. tol. at MMC) + 0.075 (bonus tol. at LMC) = 12.33 (rounded off). This is also correct. So, our reintegration of the Statistical Tolerance was successful and followed a reverse logical progression of the methods used to calculate the Statistical Tolerance to begin with. The key was to use the percentage that all tolerances were increased to (in this case) 128%.

Of course, there are other methods that can be used to reintegrate the tolerances that distribute them differently. Some try to help the more difficult to manufacture features by drawing tolerance from other features in the assembly. This allows the difficult-to-manufacture features to get more of the tolerance. But, if that was a factor, it probably should have been thought of and handled when the tolerances were being arithmetically calculated and before the calculation of the Statistical Tolerances began.

The tolerances calculated through the methods shown in this unit are identified with the following ST symbol. When both the statistical tolerance and the smaller arithmetic tolerance are shown, only those facilities using SPC controls are to be allowed the larger ST tolerance.

A feature given both an arithmetically calculated tolerance and a statistically calculated tolerance would look like this:

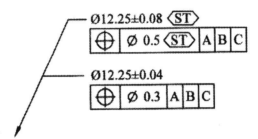

A note such as the following must be placed on the drawing: FEATURES IDENTIFIED AS STATISTICALLY TOLERANCED (ST) SHALL BE PRODUCED WITH STATISTICAL PROCESS CONTROLS, OR ELSE THEY SHALL BE HELD TO THE MORE RESTRICTIVE ARITHMETIC LIMITS.

Methods for calculating statistical tolerances vary from company to company. Some are best done by computer programs. **One such method is known as the Monte Carlo Method.** The term relates to methods that simulate manufacturing probabilities using random numbers. For example, we may use it to simulate the manufacture of dimensions on parts in an assembly. Given knowledge of manufacturing capability, random numbers are generated to simulate possible process results. After extensive averaging, one arrives at the likely amount of tolerance that will be consumed. In fact, there are a wide variety of methods that are used that are called Monte Carlo Methods. Because of this, one method and its results may be quite different than another. These methods use inferential statistics. Inferential statistics use the principle that a random sample tends to exhibit the same properties as the entire population from which it comes. **If one takes too few samples, it is possible that the results will not reflect the entire population.**

These simulations can be done with simple and available software, such as might be found in a spreadsheet program like Microsoft's Excel. A database can be set up that will simulate uniform distributions of dimensions for each variable using a random number generator. The calculated mean and standard deviation of the sample can be used to determine if an opportunity exists to increase tolerances and still have parts that meet functional requirements from a statistical standpoint.

In other words, although in a worst case analysis one would find functional difficulties with these increased tolerances (such as interferences), it is proven statistically unlikely that these difficulties will occur. In these programs, each variable can be treated differently. One is allowed to pick a distribution that is normal (given historical data about the variables) or uniform (given no historical data). Uniform distributions tend to give a more conservative answer since it is more of a guess rather than based on results given in past processing data.

Although it is never possible to achieve 100% accuracy through sampling without using the entire population, results are likely to be correct, given the statistics under the conditions we have put into place.

To Summarize Reintegrating the Statistical Tolerance for the Slot in Figures 11-6, 11-7 & 11-8

Step 1: With the slot, we take the radius of 6.095 and multiply by 2 to get 12.19.

Step 2: We then multiply the ST tolerance of 0.07 (0.055 x 1.28 = 0.07; the original tolerance on the slot radius used in the stack-up analysis of 0.055 increased to 128%) by 2 to get 0.14.

Step 3: We calculate the inner and outer boundaries by subtracting 0.14 from 12.19 = 12.05, and then adding 0.14 to 12.19 = 12.33.

Step 4: Take the original positional geometric tolerance of 0.05 at MMC and increase it to 128%. 0.05 x 1.28 = 0.064

Step 5: The original arithmetic geometric tolerance at LMC was 0.11 (the slot size tolerance of 0.06 + the position tolerance of 0.05 = 0.11). Increase it to 128% to get 0.1408.

Step 6: Add 0.064 (from step 4) to the inner boundary of 12.05 (from Step 3) to get the new MMC. The new MMC is 12.114.

Step 7: Subtract 0.1408 (from Step 5) from the outer boundary of 12.33 (from Step 3) to get the new LMC. The new LMC is 12.189 (rounded off).

Step 8: So, the new slot spec. is 12.114 – 12.189 with a position tolerance of 0.064 at MMC.

FINAL ANSWER
12.114-12.189 (ST)

| ⊕ | 0.064 (M) (ST) | A | B | C |

A note such as the following must be placed on the drawing: FEATURES IDENTIFIED AS STATISTICALLY TOLERANCED (ST) SHALL BE PRODUCED WITH STATISTICAL PROCESS CONTROLS.

A Simpler Way? Keeping the Same Size Mean Dimension instead of the Same Functional Boundary Mean

The Slot

Another much simpler approach could be used if one wanted to keep the same mean size dimension on the slot of 12.16 instead of the same mean functional boundary of 12.19. That calculation would begin by looking at the original illustration in Figure 11-6. The tolerance on the size of 12.16 of the slot is ±0.03. If we multiplied 0.03 by 1.28, we would get the new size tolerance of ±0.0384. Since the original size mean was 12.16, the new size limits would be 12.16 ± 0.0384. We would then multiply the position tolerance of 0.05 at MMC by 1.28. This would give us the new position tolerance of 0.064.

So, the new control would be:
Size: 12.16 ± 0.0384 Statistical Tolerance
(12.1216 MMC and 12.1984 LMC)

Position: 0.064 at MMC Statistical Tolerance

In this way, both the size tolerance and the position tolerance would be increased to 128%. This would generate an inner boundary of 12.1216 minus 0.064 = 12.0576 (instead of 12.05) and an outer boundary of 12.1984 LMC plus 0.064 position tolerance at MMC plus the 0.0768 bonus tolerance = 12.3392 (instead of 12.328 shown as a rounded-off 12.33). The mean of these boundaries is 12.0576 plus 12.3392 = 24.3968 divided by 2 = 12.1984 (instead of 12.19).

The tolerance difference between 12.05 and 12.328 is 0.278. The tolerance difference between 12.0576 and 12.3392 is 0.2816. So, these methods produce answers that differ only slightly in their boundary means and their increase in the tolerance on the slot. It depends on what is more important to the person doing the increase as to which method is followed.

FINAL ANSWER

12.16±0.0384 ⓢⓣ

| ⌖ | 0.064 Ⓜ ⓢⓣ | A | B | C |

A note such as the following must be placed on the drawing: FEATURES IDENTIFIED AS STATISTICALLY TOLERANCED ⓢⓣ SHALL BE PRODUCED WITH STATISTICAL PROCESS CONTROLS.

The Tab

The tab would be done the same way.

More Statistical Formulas and Symbols

The following summarizes functions that deal with population parameters, target value and upper and lower specification limits.

Cp

Fp

Fp calculates the spread of the population of measured features about the average. It is the portion of the population that fits inside of the plus or minus 3 sigma range that is centered about the average of the upper specification limit minus the lower specification limit. It is expressed by the formula:

$$Fp = \frac{U-L}{6\sigma}$$

This is a kind of actual display of what the Root Sum Square (and other statistics formulas) formula attempts to predict--something often referred to as 'natural tolerance'. **The Root Sum Square formula (among others) tries to predict the amount of tolerance *likely to be consumed* by manufacturing within the plus and minus 3 sigma range.** From the Fp, one could tell the actual amount of tolerance (deduced through part measurement) that has been consumed by manufacturing that is within the plus and minus 3 sigma range.

The Root Sum Square formula is:

$$T_A = \sqrt{T_1^2 + T_2^2 + T_3^2 + \ldots T_N^2}$$

> where T_A = tolerance predicated to be likely consumed under a Gaussian Frequency Curve distribution of features or parts
> and T_1^2 = the tolerance on the first feature squared, etc.

Cpk

Fpk

Fpk measures both the spread of the population and its deviation of the location. It determines the fraction of three sigma (half of 6 sigma) that the population, as measured, has spread (about the average), while fitting within the smaller of the specification 'half ranges', (U-µ) and (µ-L), measured from the population average. It is expressed by the formula:

Fpk = min (Fpl, Fpu), where

$$Fpl = \frac{\mu - L}{3\sigma}$$

and

$$Fpu = \frac{U - \mu}{3\sigma}$$

229

Fc

Fc is a measure of how much the average of the population has departed in location from the target value. It shows the departure of the population average μ from the target value τ in relation to the half range specification. Fc is expressed by the formula:

Fc = max (Fcl, Fcu), where

$$Fcl = \frac{\tau - \mu}{\tau - L}$$

and

$$Fcu = \frac{\mu - \tau}{U - \tau}$$

Fpm

Fpm is a measure the RMS (Root Mean Square) deviation of the population variable from the target value τ.

Fpm – Fp when $\mu = \tau$. Fpm is the portion of the population spread from the target that fits inside of the specification range. It is expressed by the formula:

$$Fpm = \frac{U - L}{6\sqrt{\sigma^2 + (\mu - \tau)^2}}$$

Glossary of Statistical Terms

Attribute Data: The term used to describe the type of "good versus bad" information given by receiver gages. Attribute data contains a lack of variables data. It does not relate how good or how bad a feature is, just the presence or absence of a characteristic. The information is related in the number of conforming versus non-conforming pieces.

Average or Arithmetic Mean: When a group of dimensions is taken, the measured values added together, then divided by the size of the group, an average or arithmetic mean is derived.

Bias: Bias in measurement is systemic error leading to a difference between the true value of the population of features being measured and the average result of measurements.

Bell-Shaped Curve: A distribution showing a central peak and a smooth, symmetric tapering off on either side, such as a Gaussian curve.

Calibration: Adjusting an instrument using a more accurate reference standard.

Capability: The ability of a gage, machine or procedure to hold a certain percentage of products within specification limits. Many companies consider a process capable when 99.73 percent or more of the features or parts being measured fall within the specification limits. Capability can be expressed in C_p, C_{pk}, C_R and other methods.

Cause and Effect Diagram: A diagram showing the relationships between all process inputs and their resulting problem(s) being investigated which effect the process.

c-Chart: A control chart showing the number of defects found in a subgroup of fixed size.

Characteristic: A geometric trait or product specification that is measured or examined to determine conformance. A dimension or parameter of a part that can be measured and then monitored for capability and control.

Control of a Process: A process is termed in statistical control when the process exhibits only random variations. When control charts are used, a state of statistical control is assumed to exist when all monitored points remain between stated control limits.

Control Chart: A representation, usually graphical in nature, used to keep track of outputs.

Control Limits: Boundaries that are statistically calculated that are used to determine if a process is in or out of statistical control. Control Limits and tolerance limits are not the same.

C_p: C_p is a capability index defined by the following formula: CP equals the tolerance divided by 6s.

Cpk: C_{pk} is a capability index that combines C_p and k. K is a measure of difference between the process mean and the specification mean (nominal). C_{pk} is used to determine if the process will produce units within the tolerance limits. C_{pk} equals the lesser of the following two formulae: **Formula 1.** (The upper specification limit minus the mean) divided by 3 times the standard deviation of a sample. **Formula 2.** (The mean minus the lower specification limit) divided by 3 times the standard deviation.

C_R: C_R is the inverse of C_p. C_R equals 6σ divided by the tolerance. The smaller the value the more capable the process.

Distribution: A display of values to show frequency of occurrence.

Mean: The numerical value in a distribution of values calculated by adding all values, then dividing by the number of values that have been added. Same as average or arithmetic mean.

Median: The value in a group of numbers that falls in the middle between the lowest and highest.

R Chart: A control chart that shows the range of variation of the individual elements of a sample.

Range: The difference between the lowest and highest values in a set of values. The range is expected to increase as the sample size and the standard deviation increases.

Reliability: The probability that a product will properly function for a period of time under certain conditions.

Repeatability: The variation in measurements that are obtained when one inspector using the same tool(s) measures the same feature characteristic of the same part.

Reproducibility: Reproducibility is the variation in the average of measurements made by a variety of operators that use the same tools measuring the same characteristics of the same parts.

Sigma: The standard deviation of a statistical population is often characterized by the Greek letter σ, which stands for sigma. However, the upper case Greek letter for sigma Σ stands for summation ($n_1 + n_2 + n_3 + n_4$, and so on).

Specifications: The required properties of a workpiece. Specifications may include the upper and lower limits of a dimension, a texture of surface finish or anything required of a feature.

Statistical Process Control: Methods of statistics used for the analysis and control of variation in a process. The use of control charts to determine significant changes in a process.

Variables Data: The measured values of a feature. Quantitative data capable of measuring (having a value) anywhere within a given range of values.

Standard Deviation: Standard deviation is a measure of the variation of the members of a statistical sample.

Standard deviation is a way of quantifying how the values in a distribution depart from the average value of the distribution. The formula that follows may be used to calculate a sample standard deviation. A sample standard deviation is the standard deviation calculated from sample data.

$$\sigma_x = \sqrt{\frac{\Sigma(X-\overline{X})^2}{(n-1)}}$$

where: σ_x = sample standard deviation

Σ = summation symbol ($n_1 + n_2 + n_3$, etc.)

X = individual readings

$\overline{X}$ = average

n = number of readings

$\sqrt{}$ = square root symbol

To use this formula to calculate the standard deviation using the sample values of 2, 4, 6, 8 10 and 12, do the following:

First calculate $\overline{X}$.

$$\overline{X} = \frac{(2+4+6+8+10+12)}{6} = \frac{42}{6} = 7$$

Then subtract $\overline{X}$ from each value, square the results and sum the squares:

$(X-\overline{X})$	$(X-\overline{X})^2$
2-7= -5	$-5^2 = 25$
4-7= -3	$-3^2 = 9$
6-7= -1	$-1^2 = 1$
8-7= 1	$1^2 = 1$
10-7= 3	$3^2 = 9$
12-7= 5	$5^2 = 25$
	SUM OF SQUARES = 70

The next step is to divide the SUM OF THE SQUARES by the number of values minus one, and then calculate the square root.

$\frac{70}{5} = 14$; $\sqrt{14} = 3.74$ = standard deviation (σ_x)

CHAPTER 11
EXERCISES

Exercise 11-1

Using the illustrations below, calculate the Statistical Tolerances for the Wall, Slot, Tab and Overall dimensions using the RSS methodology (<u>without</u> a safety/correction factor).

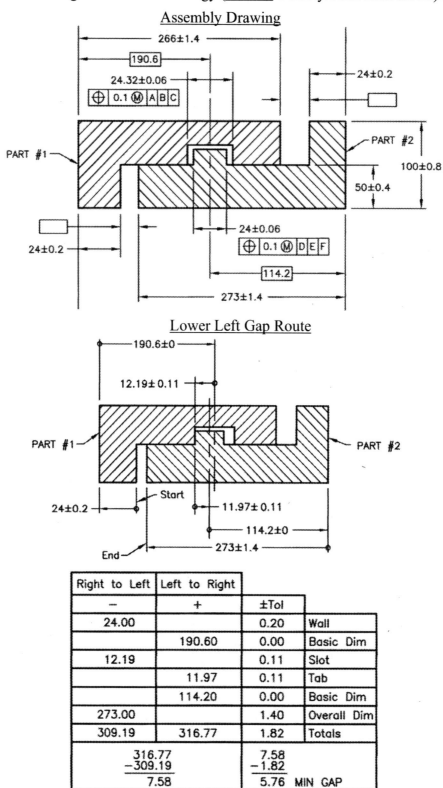

Exercise 11-2

Using the illustrations below and on the next page from Chapter 5, calculate the Statistical Tolerances to be reintegrated into the assembly for all features used in this MIN GAP calculation. Instead of using the straight RSS formula, use the following RSS formula, with a 1.5 'safety correction' factor.

$$T_A = 1.5\sqrt{T_1^2 + T_2^2 + T_3^2 + \ldots T_N^2}$$

Assembly Drawing

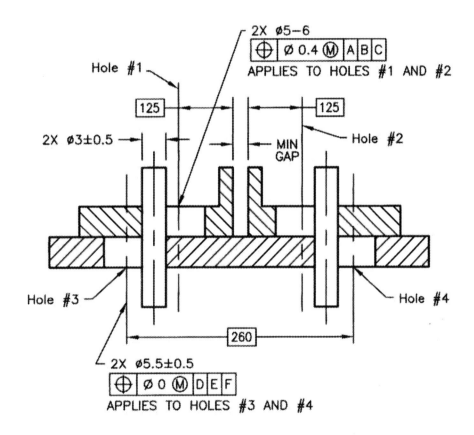

Exercise 11-2 (continued)

Minimum Gap Route

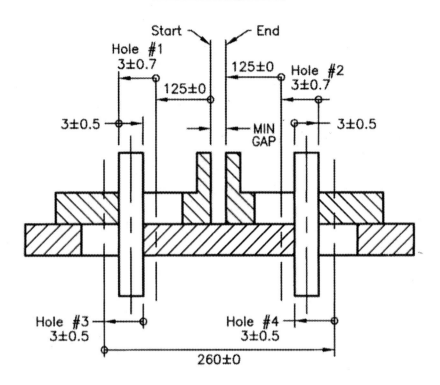

Loop Diagram

Right to Left	Left to Right	±Tol	
−	+		
125		0.0	Basic Dim
3		0.7	Hole #1
	3	0.5	Pin
3		0.5	Hole #3
	260	0.0	Basic Dim
3		0.5	Hole #4
	3	0.5	Pin
3		0.7	Hole #2
125		0.0	Basic Dim
262	266	3.4	Totals

```
266        4.0
-262      -3.4
----      ----
  4        0.6  MIN GAP
```

Tolerance Stack-Up Analysis
ANSWER SECTION

ANSWER - Exercise 1-1

1) $\phi 100^{+3}_{-1}$ $\begin{array}{r}103\\+\ 99\\\hline 202\end{array}$ $\dfrac{202}{2}=101$ & $\begin{array}{r}103\\-\ 99\\\hline 4\end{array}$ $\dfrac{4}{2}=2$ So; 101±2

2) $\phi 150-155$ $\begin{array}{r}155\\+150\\\hline 305\end{array}$ $\dfrac{305}{2}=152.5$ & $\begin{array}{r}155\\-150\\\hline 5\end{array}$ $\dfrac{5}{2}=2.5$ So; 152.5±2.5

3) $200^{+2}_{\ 0}$ $\begin{array}{r}202\\+200\\\hline 402\end{array}$ $\dfrac{402}{2}=201$ & $\begin{array}{r}202\\-200\\\hline 2\end{array}$ $\dfrac{2}{2}=1$ So; 201±1

4) $300^{+0.56}_{-0.43}$ $\begin{array}{r}300.56\\+299.57\\\hline 600.13\end{array}$ $\dfrac{600.13}{2}=300.065$ & $\begin{array}{r}300.56\\-299.57\\\hline 0.99\end{array}$ $\dfrac{0.99}{2}=0.495$ So; 300.065±0.495

5) $250.06-266.08$ $\begin{array}{r}266.08\\+250.06\\\hline 516.14\end{array}$ $\dfrac{516.14}{2}=258.07$ & $\begin{array}{r}266.08\\-250.06\\\hline 16.02\end{array}$ $\dfrac{16.02}{2}=8.01$ So; 258.07±8.01

6) $\phi 30^{\ 0}_{-0.47}$ $\begin{array}{r}30.00\\+29.53\\\hline 59.53\end{array}$ $\dfrac{59.53}{2}=29.765$ & $\begin{array}{r}30.00\\-29.53\\\hline 0.47\end{array}$ $\dfrac{0.47}{2}=0.235$ So; 29.765±0.235

7) $500^{+0.26}_{-0.37}$ $\begin{array}{r}500.26\\+499.63\\\hline 999.89\end{array}$ $\dfrac{999.89}{2}=499.945$ & $\begin{array}{r}500.26\\-499.63\\\hline 0.63\end{array}$ $\dfrac{0.63}{2}=0.315$ So; 499.945±0.315

8) $25.02-25.84$ $\begin{array}{r}25.02\\+25.84\\\hline 50.86\end{array}$ $\dfrac{50.86}{2}=25.43$ & $\begin{array}{r}25.84\\-25.02\\\hline 0.82\end{array}$ $\dfrac{0.82}{2}=0.41$ So; 25.43±0.41

9) $\phi 8.66-8.90$ $\begin{array}{r}8.90\\+8.66\\\hline 17.56\end{array}$ $\dfrac{17.56}{2}=8.78$ & $\begin{array}{r}8.90\\-8.66\\\hline 0.24\end{array}$ $\dfrac{0.24}{2}=0.12$ So; 8.78±0.12

10) $\phi 4.21^{+0.25}_{-0.36}$ $\begin{array}{r}4.46\\+3.85\\\hline 8.31\end{array}$ $\dfrac{8.31}{2}=4.155$ & $\begin{array}{r}4.46\\-3.85\\\hline 0.61\end{array}$ $\dfrac{0.61}{2}=0.305$ So; 4.155±0.305

ANSWER - Exercise 1-2

		Outer Boundary	Inner Boundary
1)	⌀20±2 Hole ⊕ ⌀2 Ⓜ A B C	22+6=28	18−2=16
2)	⌀20±2 Hole ⊕ ⌀2 Ⓛ A B C	22+2=24	18−6=12
3)	⌀20±2 Hole ⊕ ⌀2 A B C	22+2=24	18−2=16
4)	⌀20 $^{+3}_{-2}$ Shaft ⊕ ⌀0.5 Ⓜ A B C	23.0+0.5=23.5	18.0−5.5=12.5
5)	⌀15−16 Shaft ⊕ ⌀0.4 Ⓛ A B C	16.0+1.4=17.4	15.0−0.4=14.6
6)	⌀30 $^{+0.4}_{-0.2}$ Shaft ⊕ ⌀0.3 A B C	30.4+0.3=30.7	29.8−0.3=29.5
7)	28.5 $^{+0.2}_{-0.1}$ Slot ⊕ 0.8 Ⓜ X Y Z	28.7+1.1=29.8	28.4−0.8=27.6
8)	23.08−24.04 Tab ⊕ 0.5 Ⓜ T E C	24.04+0.50=24.54	23.08−1.46=21.62
9)	75±2 Slot ⊕ 1 Ⓛ C A B	77+1=78	73−5=68
10)	70±0.5 Tab ⊕ 0.5 Ⓛ D E F	70.5+1.5=72	69.5−0.5=69
11)	12−13 Slot ⊕ 0.3 G A F	13+0.3=13.3	12−0.3=11.7
12)	9.5−10.6 Tab ⊕ 1.2 H J K	10.6+1.2=11.8	9.5−1.2=8.3

ANSWER - Exercise 1-2 (continued)

①
```
  28 = O.B.        28 = O.B.
+ 16 = I.B.      − 16 = I.B.
──────────       ──────────
  44 = Sum         12 = Diff.
```

$$\frac{44}{2} = 22 = \text{Mean} \quad \& \quad \frac{12}{2} = 6 = \pm\text{Tol.}$$

So; 22±6

②
$$\begin{array}{ll} 24 = \text{O.B.} & 24 = \text{O.B.} \\ +\,12 = \text{I.B.} & -\,12 = \text{I.B.} \\ \hline 36 = \text{Sum} & 12 = \text{Diff.} \end{array}$$

$\dfrac{36}{2} = 18 = \text{Mean}$ & $\dfrac{12}{2} = 6 = \pm\text{Tol.}$

So; 18±6

③
$$\begin{array}{ll} 24 = \text{O.B.} & 24 = \text{O.B.} \\ +\,16 = \text{I.B.} & -\,16 = \text{I.B.} \\ \hline 40 = \text{Sum} & 8 = \text{Diff.} \end{array}$$

$\dfrac{40}{2} = 20 = \text{Mean}$ & $\dfrac{8}{2} = 4 = \pm\text{Tol.}$

So; 20±4

④
$$\begin{array}{ll} 23.5 = \text{O.B.} & 23.5 = \text{O.B.} \\ +\,12.5 = \text{I.B.} & -\,12.5 = \text{I.B.} \\ \hline 36 = \text{Sum} & 11 = \text{Diff.} \end{array}$$

$\dfrac{36}{2} = 18 = \text{Mean}$ & $\dfrac{11}{2} = 5.5 = \pm\text{Tol.}$

So; 18±5.5

⑤
$$\begin{array}{ll} 17.4 = \text{O.B.} & 17.4 = \text{O.B.} \\ +\,14.6 = \text{I.B.} & -\,14.6 = \text{I.B.} \\ \hline 32 = \text{Sum} & 2.8 = \text{Diff.} \end{array}$$

$\dfrac{32}{2} = 16 = \text{Mean}$ & $\dfrac{2.8}{2} = 1.4 = \pm\text{Tol.}$

So; 16±1.4

⑥
$$\begin{array}{ll} 30.7 = \text{O.B.} & 30.7 = \text{O.B.} \\ +\,29.5 = \text{I.B.} & -\,29.5 = \text{I.B.} \\ \hline 60.2 = \text{Sum} & 1.2 = \text{Diff.} \end{array}$$

$\dfrac{60.2}{2} = 30.1 = \text{Mean}$ & $\dfrac{1.2}{2} = 0.6 = \pm\text{Tol.}$

So; 30.1±0.6

⑦
$$\begin{array}{r} 29.8 = \text{O.B.} \\ + 27.6 = \text{I.B.} \\ \hline 57.4 = \text{Sum} \end{array}$$
$$\begin{array}{r} 29.8 = \text{O.B.} \\ - 27.6 = \text{I.B.} \\ \hline 2.2 = \text{Diff.} \end{array}$$

$\dfrac{57.4}{2} = 28.7 = \text{Mean}$ & $\dfrac{2.2}{2} = 1.1 = \pm \text{Tol.}$

So; 28.7 ± 1.1

⑧
$$\begin{array}{r} 24.54 = \text{O.B.} \\ + 21.62 = \text{I.B.} \\ \hline 46.16 = \text{Sum} \end{array}$$
$$\begin{array}{r} 24.54 = \text{O.B.} \\ - 21.62 = \text{I.B.} \\ \hline 2.92 = \text{Diff.} \end{array}$$

$\dfrac{46.16}{2} = 23.08 = \text{Mean}$ & $\dfrac{2.92}{2} = 1.46 = \pm \text{Tol.}$

So; 23.08 ± 1.46

⑨
$$\begin{array}{r} 78 = \text{O.B.} \\ + 68 = \text{I.B.} \\ \hline 146 = \text{Sum} \end{array}$$
$$\begin{array}{r} 78 = \text{O.B.} \\ - 68 = \text{I.B.} \\ \hline 10 = \text{Diff.} \end{array}$$

$\dfrac{146}{2} = 73 = \text{Mean}$ & $\dfrac{10}{2} = 5 = \pm \text{Tol.}$

So; 73 ± 5

⑩
$$\begin{array}{r} 72 = \text{O.B.} \\ + 69 = \text{I.B.} \\ \hline 141 = \text{Sum} \end{array}$$
$$\begin{array}{r} 72 = \text{O.B.} \\ - 69 = \text{I.B.} \\ \hline 3 = \text{Diff.} \end{array}$$

$\dfrac{141}{2} = 70.5 = \text{Mean}$ & $\dfrac{3}{2} = 1.5 = \pm \text{Tol.}$

So; 70.5 ± 1.5

⑪
$$\begin{array}{r} 13.3 = \text{O.B.} \\ + 11.7 = \text{I.B.} \\ \hline 25 = \text{Sum} \end{array}$$
$$\begin{array}{r} 13.3 = \text{O.B.} \\ - 11.7 = \text{I.B.} \\ \hline 1.6 = \text{Diff.} \end{array}$$

$\dfrac{25}{2} = 12.5 = \text{Mean}$ & $\dfrac{1.6}{2} = 0.8 = \pm \text{Tol.}$

So; 12.5 ± 0.8

⑫
$$\begin{array}{r} 11.8 = \text{O.B.} \\ + 8.3 = \text{I.B.} \\ \hline 20.1 = \text{Sum} \end{array}$$
$$\begin{array}{r} 11.8 = \text{O.B.} \\ - 8.3 = \text{I.B.} \\ \hline 3.5 = \text{Diff.} \end{array}$$

$\dfrac{20.1}{2} = 10.05 = \text{Mean}$ & $\dfrac{3.5}{2} = 1.75 = \pm \text{Tol.}$

So; 10.05 ± 1.75

ANSWER - Exercise 2-1, Worksheet #1

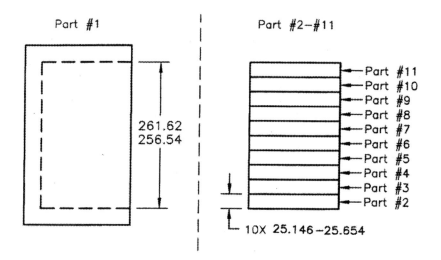

```
  261.62      261.62
+ 256.54    − 256.54
  ──────      ──────
  518.16        5.08

1/2 of 518.16 = 259.08
1/2 of 5.08 = 2.54

So, expressed as an equal
bilaterally toleranced
dimension: 259.08 ± 2.54
```

```
 25.654     25.146     256.54     256.54
 X 10       X 10     + 251.46   − 251.46
 ─────      ─────     ──────     ──────
 256.54     251.46      508        5.08

1/2 of 508 = 254
1/2 of 5.08 = 2.54

So, expressed as an equal bilaterally
toleranced dimension: 254 ± 2.54

or:
  25.654     25.654     50.8/2 = 25.4
+ 25.146   − 25.146    0.508/2 = 0.254
  ──────     ──────
   50.8       0.508

So: 25.4 ± 0.254 ----- 10 times
```

ANSWER - Exercise 2-1, Worksheet #2

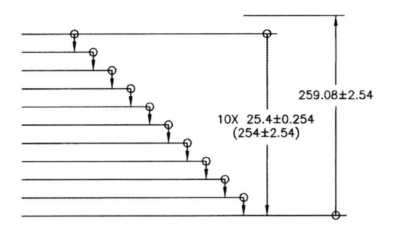

DIMENSIONS		±TOL
−	+	
254		2.54
	259.08	2.54
254	259.08	5.08 TOTALS
259.08 − 254.00 5.08	5.08 + 5.08 10.16=MAX GAP	5.08 − 5.08 0=MIN GAP (zero interference)

Possible solutions to interference problems, should ever they occur:
1. Increase MMC size of cavity.
2. Decrease MMC size of plates.
3. Decrease tolerance on cavity (effectively increasing the MMC).
4. Decrease the tolerances on the plates (effectively decreasing their collective MMC).

ANSWER - Exercise 3-1, Worksheet #1

Left to Right

```
 40.53      79.85      40.53
+39.32        2       -39.32
 79.85    is 39.925    1.21=±0.605

 38.71      77.11      38.71
+38.40        2       -38.40
 77.11    is 38.555    0.31=±0.155
```

Bottom to Top

```
 40.53      79.85      40.53
+39.32        2       -39.32
 79.85    is 39.925    1.21=±0.605

 37.50      73.17      37.50
+35.67        2       -35.67
 73.17    is 36.585    1.83=±0.915
```

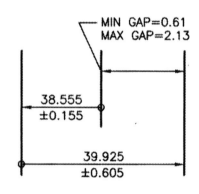

MIN GAP=0.61
MAX GAP=2.13

38.555 ±0.155
39.925 ±0.605

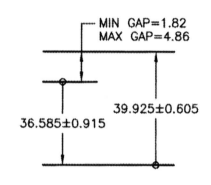

MIN GAP=1.82
MAX GAP=4.86

36.585±0.915
39.925±0.605

DIMENSIONS		TOLERANCE
−	+	±
38.555		0.155
	39.925	0.605
38.555	39.925	0.760 TOTALS

```
  39.925
 -38.555
   1.37

1.37+0.76=2.13 MAX GAP
1.37-0.76=0.61 MIN GAP
```

DIMENSIONS		TOLERANCE
−	+	±
36.585		0.915
	39.925	0.605
36.585	39.925	1.520 TOTALS

```
  39.925
 -36.585
   3.34

3.34+1.52=4.86 MAX GAP
3.34-1.52=1.82 MIN GAP
```

ANSWER - Exercise 4-1, Worksheet #1 (Pertinent Numbers for Both Gap Calculations)

```
   52.35    52.35       32.12    32.12       25.00    25.00
 + 47.63   - 47.63     + 29.76  - 29.76     + 22.64  - 22.64
   99.98     4.72        61.88     2.36       47.64     2.36
```

99.98/2=49.99   61.88/2=30.94   47.64/2=23.82

4.72/2=2.36   2.36/2=1.18   2.36/2=1.18

49.99 ± 2.36   30.94 ± 1.18   23.82 ± 1.18

```
             39.32    39.32        38.10    38.10
           + 38.12   - 38.12      + 36.89  - 36.89
             77.44     1.20         74.99     1.21
```

77.44/2=38.72   74.99/2=37.495

1.2/2=0.6   1.21/2=0.605

38.72 ± 0.6   37.495 ± 0.605

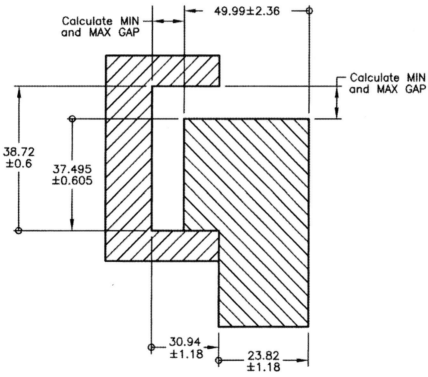

DIMENSIONS		TOL
−37.495		0.605
	+38.720	0.600
+1.225 DIM DIFF		1.205 TOTALS
1.225		1.225
+1.205		−1.205
2.43=MAX GAP		0.02=MIN GAP

DIMENSIONS		TOL
+30.940		1.180
+23.820		1.180
	−49.990	2.360
+54.760	−49.990	4.720 TOTALS

```
    54.76
  − 49.99
    4.77
  DIM DIFF
  4.77    4.77
+ 4.72  − 4.72
  9.49    0.05
  MAX     MIN
  GAP     GAP
```

ANSWER - Exercise 5-1, Worksheet #1

Resultant Condition of Hole #1= ⌀8.5 + 1.4 = ⌀9.9
Virtual Condition of Hole #1= ⌀7.5 - 0.4 = ⌀7.1

Resultant Condition of Hole #2= ⌀8.5 + 1.4 = ⌀9.9
Virtual Condition of Hole #2= ⌀7.5 - 0.4 = ⌀7.1

Resultant Condition of Hole #3= ⌀8.65 + 0.81 = ⌀9.46
Virtual Condition of Hole #3= ⌀7.85 - 0.01 = ⌀7.84

Resultant Condition of Hole #4= ⌀8.65 + 0.81 = ⌀9.46
Virtual Condition of Hole #4= ⌀7.85 - 0.01 = ⌀7.84

Res. Cond. Hole #1= ⌀9.9	Res. Cond. Hole #1= ⌀9.9
+ Virt. Cond. Hole #1= ⌀7.1	- Virt. Cond. Hole#1= ⌀7.1
Sum 17	Difference 2.8

Sum and Difference Divided by 2= ⌀8.5±1.4=Mean Dia. ± Mean Tol.
Mean Dia. ± Mean Tol. Divided by 2=R4.25±0.7=Mean Radius ± Mean Tol.

Res. Cond. Hole #2= ⌀9.9	Res. Cond. Hole #2= ⌀9.9
+ Virt. Cond. Hole #2= ⌀7.1	- Virt. Cond. Hole#2= ⌀7.1
Sum 17	Difference 2.8

Sum and Difference Divided by 2= ⌀8.5±1.4=Mean Dia. ± Mean Tol.
Mean Dia. ± Mean Tol. Divided by 2=R4.25±0.7=Mean Radius ± Mean Tol.

Res. Cond. Hole #3= ⌀9.46	Res. Cond. Hole #3= ⌀9.46
+ Virt. Cond. Hole #3= ⌀7.84	- Virt. Cond.Hole #3= ⌀7.84
Sum 17.3	Difference 1.62

Sum and Difference Divided by 2= ⌀8.65±0.81=Mean Dia. ± Mean Tol.
Mean Dia. ± Mean Tol. Divided by 2=R4.325±0.405=Mean Radius ± Mean Tol.

Res. Cond. Hole #3= ⌀9.46	Res. Cond. Hole #3= ⌀9.46
+ Virt. Cond. Hole #3= ⌀7.84	- Virt. Cond.Hole #3= ⌀7.84
Sum 17.3	Difference 1.62

Sum and Difference Divided by 2= ⌀8.65±0.81=Mean Dia. ± Mean Tol.
Mean Dia. ± Mean Tol. Divided by 2=R4.325±0.405=Mean Radius ± Mean Tol.

ANSWER - Exercise 5-1, Worksheet #2

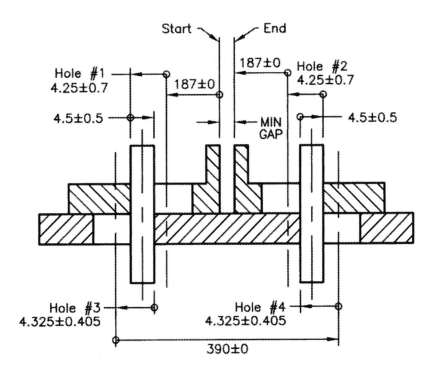

Loop Diagram

Right to Left	Left to Right	±Tol	
−	+		
187.000		0.000	Basic Dim
4.250		0.700	Hole #1
	4.500	0.500	Pin
4.325		0.405	Hole #3
	390.000	0.000	Basic Dim
4.325		0.405	Hole #4
	4.500	0.500	Pin
4.250		0.700	Hole #2
187.000		0.000	Basic Dim
391.150	399.000	3.210	Totals
399.00 −391.15 7.85		7.85 −3.21 4.64 MIN GAP	

ANSWER - Exercise 5-1, Worksheet #3

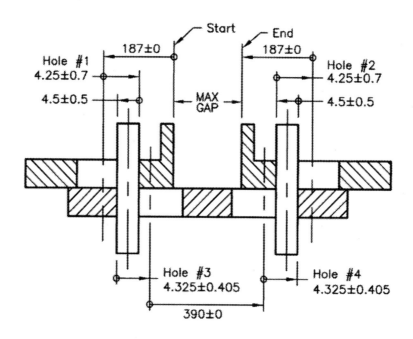

Loop Diagram

Right to Left	Left to Right	±Tol	
−	+		
187.000		0.000	Basic Dim
	4.250	0.700	Hole #1
4.500		0.500	Pin
	4.325	0.405	Hole #3
	390.000	0.000	Basic Dim
	4.325	0.405	Hole #4
4.500		0.500	Pin
	4.250	0.700	Hole #2
187.000		0.000	Basic Dim
383.000	407.150	3.210	Totals

```
  407.15         24.15
 −383.00         +3.21
  ─────         ──────
   24.15        27.36 MAX GAP
```

ANSWER - Exercise 6-1, Worksheet #1

Virtual and Resultant Condition Calculations for Tolerance Stack-Up Analysis for Slot and Tab

Resultant Condition of Slot = 24.38 + 0.22 = 24.6

Virtual Condition of Slot = 24.26 - 0.10 = 24.16

Resultant Condition of Tab = 23.94 - 0.22 = 23.72

Virtual Condition of Tab = 24.06 + 0.1 = 24.16

Res. Cond. Slot = 24.60 − Virt. Cond. Slot = 24.16 Difference = 0.44	Virt. Cond. Tab = 24.16 − Res. Cond. Tab = 23.72 Difference = 0.44
1/2 Difference Slot = 0.22	1/2 Difference Tab = 0.22
Resultant Condition Slot = 24.60 + Virtual Condition Slot = 24.16 Sum = 48.76	Virtual Condition Tab = 24.16 + Resultant Condition Tab = 23.72 Sum = 47.88
1/2 Sum of R.C. and V.C. of Slot = 24.38	1/2 Sum of V.C. and R.C. of Tab = 23.94
1/2 Sum ± 1/2 Diff. of Slot = 24.38 ± 0.22	1/2 Sum ± 1/2 Diff. of Tab = 23.94 ± 0.22
1/2 of 1/2 Sum ± 1/2 of 1/2 Diff. Slot = 12.19 ± 0.11	
	1/2 of 1/2 Sum ± 1/2 of 1/2 Diff. Tab = 11.97 ± 0.11

ANSWER - Exercise 6-1, Worksheet #2

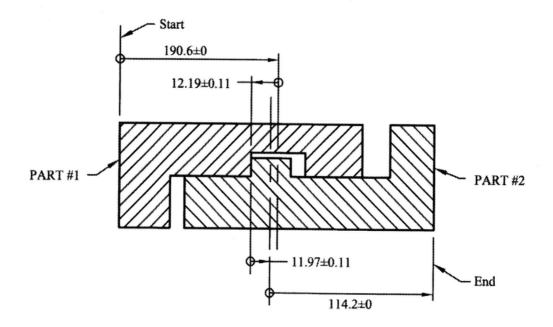

Right to Left	Left to Right		
-	+	±Tol	
	190.60	0.00	Basic Dim
12.19		0.11	Slot
	11.97	0.11	Tab
	114.20	0.00	Basic Dim
12.19	316.77	0.22	Totals

```
316.77          304.58
- 12.19        -  0.22
 304.58         304.36   MIN Overall Measurement
```

ANSWER - Exercise 6-1, Worksheet #3

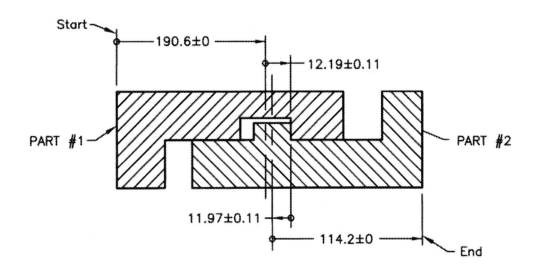

Right to Left	Left to Right		
−	+	±Tol	
	190.60	0.00	Basic Dim
	12.19	0.11	Slot
11.97		0.11	Tab
	114.20	0.00	Basic Dim
11.97	316.99	0.22	Totals

```
 316.99       305.02
- 11.97      +  0.22
 305.02       305.24   MAX Overall Measurement
```

ANSWER - Exercise 6-1, Worksheet #4

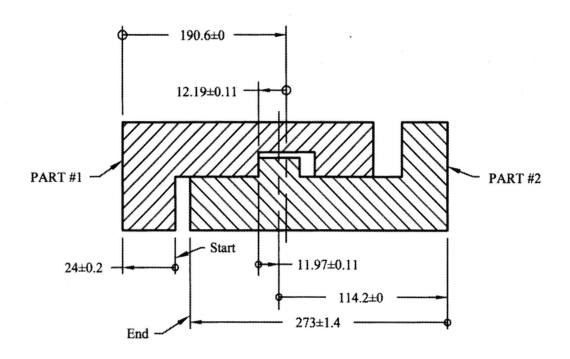

Right to Left	Left to Right	±Tol	
-	+	±Tol	
24.00		0.20	Wall
	190.60	0.00	Basic Dim
12.19		0.11	Slot
	11.97	0.11	Tab
	114.20	0.00	Basic Dim
273.00		1.40	Overall Dim
309.19	316.77	1.82	Totals

```
  316.77          7.58
- 309.19        - 1.82
  ─────          ─────
    7.58          5.76   MIN GAP
```

ANSWER - Exercise 6-1, Worksheet #5

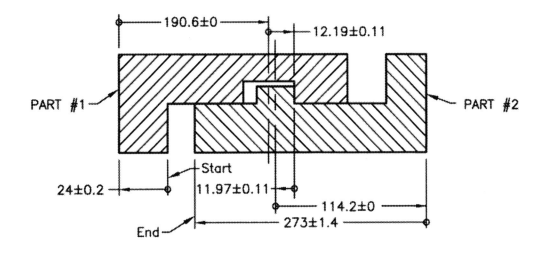

Right to Left	Left to Right		
−	+	±Tol	
24.00		0.20	Wall
	190.60	0.00	Basic Dim
	12.19	0.11	Slot
11.97		0.11	Tab
	114.20	0.00	Basic Dim
273.00		1.40	Overall Dim
308.97	316.99	1.82	Totals

```
  316.99         8.02
 -308.97        +1.82
 ───────        ─────
    8.02         9.84  MAX GAP
```

ANSWER - Exercise 6-1, Worksheet #6

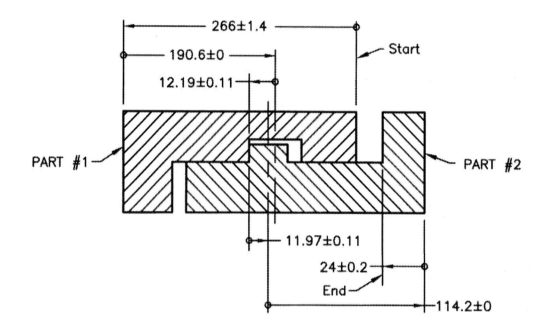

Right to Left	Left to Right	±Tol	
−	+		
266.00		1.40	Overall
	190.60	0.00	Basic Dim
12.19		0.11	Slot
	11.97	0.11	Tab
	114.20	0.00	Basic Dim
24.00		0.20	Wall
302.19	316.77	1.82	Totals
316.77 − 302.19 = 14.58		14.58 − 1.82 = 12.76 = MIN GAP	

ANSWER - Exercise 6-1, Worksheet #7

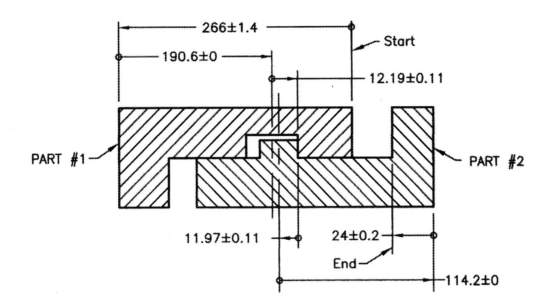

Right to Left	Left to Right	±Tol	
−	+		
266.00		1.40	Overall
	190.60	0.00	Basic Dim
	12.19	0.11	Slot
11.97		0.11	Tab
	114.20	0.00	Basic Dim
24.00		0.20	Wall
301.97	316.99	1.82	Totals

```
  316.99        15.02
 -301.97      + 1.82
  ──────      ──────
   15.02       16.84 = MAX GAP
```

ANSWER - Exercise 6-2

Step 1:

      250.2 = MMC
  +   0.2 = Geometric Tolerance at MMC
    Ø250.4 = Outer Boundary of Shaft

Ø250.4/2 = Radius of 125.2

Step 2:

Coupling Datum Feature B LMC = Ø59.95

Ø59.95/2 = Radius of 29.975

Step 3:

Crankshaft Datum Feature D LMC = Ø60.12

Ø60.12/2 = Radius of 30.06

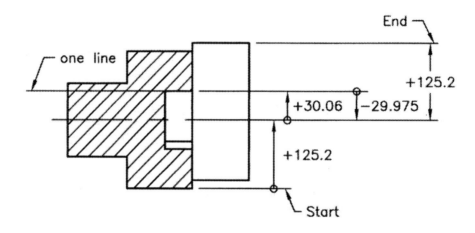

−	+
	125.200
	30.060
29.975	
	125.200
29.975	280.460	Totals

    280.460
−   29.975
    250.485 = MAX Overall Dimension

Although the MAX overall dimension is 250.485 while stationary, the maximum overall housing requirement would be 0.085 larger while rotating at 250.485 + 0.085 (30.060 − 29.975) = <u>250.57</u>.

ANSWER - Exercises 7-1 through 7-3, Worksheet #1

Outer Boundary of Slot in Rail = 38.1 + 0.2 = 38.3

Inner Boundary of Slot in Rail = 37.9 − 0.2 = 37.7

Inner Boundary of Width of Block = 35.9 − 0.2 = 35.7

Outer Boundary of Width of Block = 36.1 + 0.2 = 36.3

Outer Boundary Slot = 38.3 − Inner Boundary Slot = 37.7 Difference = 0.6 1/2 Difference of Slot = 0.3 Outer Boundary Slot = 38.3 + Inner Boundary Slot = 37.7 Sum = 76 1/2 Sum of O.B. and I.B. of Slot = 38 1/2 Sum ±1/2 Diff. of Slot = 38±0.3 1/2 of 1/2 Sum ±1/2 of 1/2 Diff. Slot = 19±0.15	Outer Boundary Block = 36.3 − Inner Boundary Block = 35.7 Difference = 0.6 1/2 Difference of Block = 0.3 Outer Boundary Block = 36.3 + Inner Boundary Block = 35.7 Sum = 72 1/2 Sum of O.B. and I.B. of Block = 36 1/2 Sum ±1/2 Diff. of Block = 36±0.3 1/2 of 1/2 Sum ±1/2 of 1/2 Diff. Block = 18±0.15

ANSWER - Exercises 7-1 through 7-3, Worksheet #2

Inner Boundary of Screw Mounted in Rail = ⌀7.46 [7.76(LMC)−0.3(Geo. Tol.)=⌀7.46]

Outer Boundary of Screw Mounted in Rail = ⌀8.3 [8.0(MMC)+0.3(Geo.Tol.)=⌀8.3]

Outer Boundary of Hole in Block = ⌀9.44

Inner Boundary of Hole in Block = ⌀8.36

Outer Boundary Mounted Screw = 8.30 − Inner Boundary Mounted Screw = 7.46 Difference = 0.84	Outer Boundary Hole = 9.44 − Inner Boundary Hole = 8.36 Difference = 1.08
1/2 Difference Mounted Screw = 0.42	1/2 Difference Hole = 0.54
Outer Boundary Mounted Screw = 8.30 + Inner Boundary Mounted Screw = 7.46 Sum = 15.76	Outer Boundary Hole = 9.44 + Inner Boundary Hole = 8.36 Sum = 17.80
1/2 Sum of O.B. and I.B. Screw = 7.88	1/2 Sum of O.B. and I.B. Hole = 8.90
1/2 Sum ±1/2 Diff. Screw = 7.88±0.42	1/2 Sum ±1/2 Diff. Hole = 8.90±0.54
1/2 of 1/2 Sum ±1/2 of 1/2 Diff. Screw = 3.94±0.21	1/2 of 1/2 Sum ±1/2 of 1/2 Diff. Hole = 4.45±0.27

ANSWER - Exercise 7-4, Worksheet #1

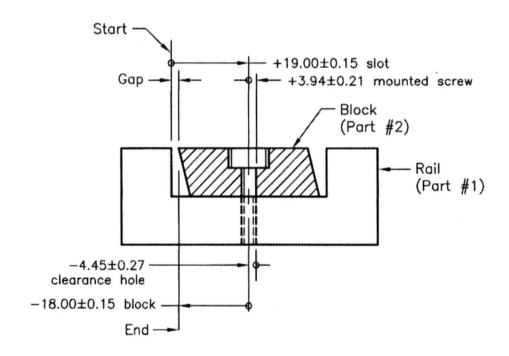

ANSWER - Exercise 7-4, Worksheet #1 (continued)

MIN GAP

+18.85 (19.00−0.15) plus +3.73 (3.94−0.21) = +22.58

−4.72 (4.45+0.27) plus −18.15 (18.00+0.15) = 22.87

+22.58 plus −22.87 = −0.29 MIN GAP

So, the maximum interference is 0.29

Or:

MIN GAP

−	+	Tolerance	
	19.00	0.15	
	3.94	0.21	
4.45		0.27	
18.00		0.15	
22.45	22.94	0.78	Totals

```
22.94    0.49
-22.45   -0.78
 0.49   -0.29 = Interference Maximum
```

ANSWER - Exercise 7-5, Worksheet #1

1.   8.0   =   MMC of Screw
       + 0.3   =   Geo. Tolerance for Threaded Holes
       Ø8.3   =   Virtual Condition of Screw (while mounted in the threaded hole)

   Ø8.3 ÷ 2   =   4.15 (1/2 the Virtual Condition Screw)

2.   37.9   =   MMC Cavity
      − 0.2   =   Geo. Tolerance for Cavity
       37.7   =   Inner Boundary of Cavity

   37.7 ÷ 2   =   18.85 (1/2 the Inner Boundary of Cavity)

3.   18.85   =   1/2 the Inner Boundary of Cavity
      − 4.15   =   1/2 the Virtual Condition Screw
       14.70   =   Minimum Airspace (between the screw surface and the cavity wall)

4.   8.66   =   MMC Clearance Hole
     − 0.30   =   Geo. Tolerance for Clearance Hole
       Ø8.36   =   Virtual Condition of Clearance Hole

   Ø8.36 ÷ 2   =   4.18 (1/2 the Virtual Condition Clearance Hole)

5.   36.1   = MMC Block
   + 0.2   = Geo. Tolerance Block
     36.3   = Outer Boundary of Block

   36.3 ÷ 2 = 18.15 (1/2 the Outer Boundary Block)

6.   18.15   = 1/2 the Outer Boundary Block
   - 4.18   = 1/2 the Virtual Condition Clearance Hole
     13.97 = Maximum Wall Thickness (between the surface of the hole to the outside edge of the block)

7.   14.70   = Minimum Airspace
   - 13.97 = Maximum Wall Thickness
     0.73   = Clearance between Rail and Block per side

Therefore, there is no interference necessary when putting these parts together - if the airspace between the screw and the clearance hole is used to adjust the parts to an optimum position for assembly.

ANSWER - Exercise 7-6, Worksheet #1

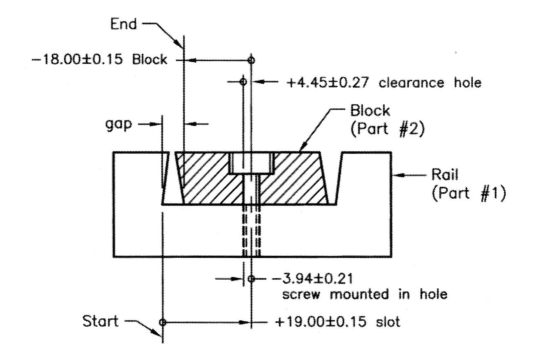

ANSWER - Exercise 7-6, Worksheet #1 (continued)

MAX GAP

−	+	Tolerance
	19.00	0.15
3.94		0.21
	4.45	0.27
18.00		0.15
21.94	23.45	0.78 Totals

```
  23.45      1.51
− 21.94    + 0.78
  1.51      2.29  =  MAX GAP
```

Or:

+19.15 (19+0.15) plus +4.72 (4.45+0.27) = +23.87

−17.85 (18−0.15) plus −3.73 (3.94−0.21) = 21.58

So: 23.87−21.58 = 2.29 MAX GAP

[This answer will be proven incorrect in subsequent worksheets.]

ALTERNATE ANSWER - Exercise 7-7, Worksheet #1 (Without Block and Slot Perpendicularity)

Step 1:
Inner boundary of the mounted screw is Ø7.46 (7.76 LMC − 0.30 Geo. Tol.).
$$\frac{\emptyset 7.46}{2} = R3.73$$

Step 2:
Outer boundary of the clearance hole is Ø9.44 (8.90 LMC + 0.54 Geo. Tol.).
$$\frac{\emptyset 9.44}{2} = R4.72$$

Step 3:
The LMC of the cavity is 38.1.
$$\frac{38.1}{2} = R19.05$$

Step 4:
The LMC of the block is 35.9.
$$\frac{35.9}{2} = R17.95$$

Now we can proceed to calculate the maximum gap.

ALTERNATE ANSWER (continued) - Exercise 7-7, Worksheet #1

MAX GAP Without Perpendicularity for Block and Cavity

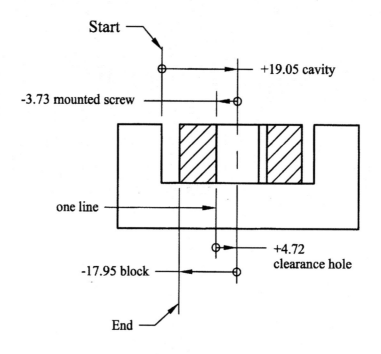

-	+	Item
	19.05	cavity
3.73		mounting screw
	4.72	clearance hole
17.95		block
21.68	23.77	Totals

23.77
- 21.68

2.09 = MAX GAP

This answer is correct without perpendicularity. The holes and screws limit the MAX GAP from being larger, unless the block and cavity are out-of- perpendicularity.

ANSWER - Exercise 7-8, Worksheet #1-MAX GAP Shortest Route

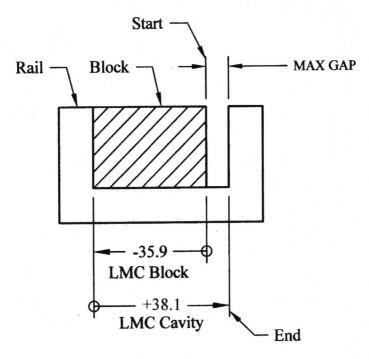

[This answer is correct.]

   38.1=LMC Cavity
- 35.9=LMC Block
    2.2=MAX GAP possible with these two mating parts using the shortest route. Page 262 shows the MAX GAP at 2.29, which the illustration above proves cannot be achieved. However, looking at all the data collected, with the holes and screws added back into the equation, the MAX GAP could be 2.2 at the widest part of the GAP when out-of-perpendicularity. The holes and screws, when added, cannot produce a MAX GAP larger than shown above, but can equal it.

ANSWER - Exercise 8-1, Problem #1

[Single Part Analysis MIN and MAX GAP Left Edge]

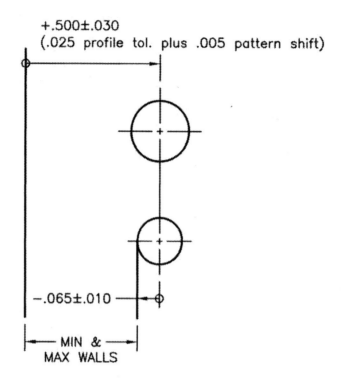

–	+	± Tolerance
	.500	.030
.065		.010
.065	.500	.040	Totals

```
   .500           .435              .435
  −.065          +.040             −.040
   ────           ─────             ─────
   .435           .475=MAX WALL     .395=MIN WALL
```

.475 MAX + .065 (1/2 hole's LMC) = .540 MAX distance from hole axis to left edge of part.

.395 MIN + .065 (1/2 hole's LMC) = .460 MIN distance from hole axis to left edge of part.

ANSWER - Exercise 8-1, Problem #2

[Single Part Analysis MIN and MAX GAP Bottom Edge]

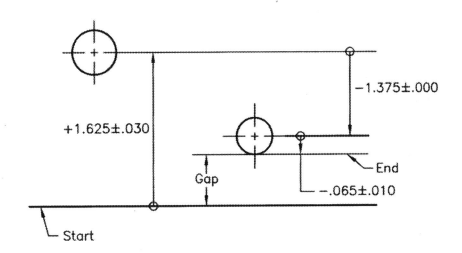

−	+	±Tol	
	1.625	.030	
1.375		.000	
.065		.010	
1.440	1.625	.040	Totals

```
  1.625        .185                .185
- 1.440      + .040              - .040
  -----       -----               -----
   .185        .225=MAX WALL       .145=MIN WALL
```

ANSWER - Exercise 8-2, Problem #1

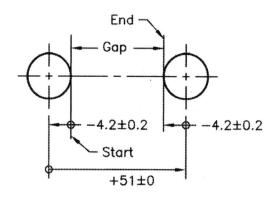

Inner Boundary of Hole = 8.0 (MMC) - 0 (Geo. Tol. at MMC) = Ø8.0
Outer Boundary of Hole = 8.4 (LMC) +0.4 (Geo. Tol. at LMC) = Ø8.8

```
  8.8 = O.B. of Hole          8.8 = O.B. of Hole
- 8.0 = I.B. of Hole        + 8.0 = I.B. of Hole
  ─────────────────           ─────────────────
  0.8 = Difference           16.8 = Sum
```

$\dfrac{0.8}{2} = 0.4 = \pm\text{Tolerance}$ $\dfrac{16.8}{2} = 8.4 = \text{Mean Dimension}$

Divide both by 2 to convert to radii = 4.2±0.2

-	+	±Tol	
4.2		0.2	
	51	0.0	
4.2		0.2	
8.4	51	0.4	Totals

```
  +51.0          42.6              42.6
+  -8.4        -  0.4            +  0.4
  ─────         ─────             ─────
  +42.6         42.2 = MIN GAP    43.0 = MAX GAP
```

267

ANSWER - Exercise 8-2, Problem #2

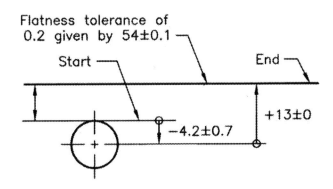

Ø8.4 (LMC of Hole) + 1.4 (Geo. Tol. at LMC to D) = Ø9.8 = O.B. Hole
Ø8.0 (MMC of Hole) - 1.0 (Geo. Tol. at MMC to D) = Ø7.0 = I.B. Hole

```
  9.8 = O.B. of Hole          9.8 = O.B. of Hole
- 7.0 = I.B. of Hole        + 7.0 = I.B. of Hole
  2.8 = Difference            16.8 = Sum
```

$\dfrac{2.8}{2} = 1.4 = \pm\text{Tolerance}$ $\dfrac{16.8}{2} = 8.4 = \text{Mean Dimension}$

Divide both by 2 to convert to radii = 4.2±0.7

```
  13.0        8.8              8.8
-  4.2      + 0.7            - 0.7
  ────       ────             ────
   8.8        9.5 = MAX WALL   8.1 = MIN WALL to Datum Plane D
                             - 0.2 = Flatness Tol. of Surface D (Rule #1)
                               ────
                               7.9 = MIN WALL to Surface D (Low Points)
```

ANSWER - Exercise 8-2, Problem #3 - Although not part of Problem #3, the Gap between the right edge and one of the holes nearest to it is calculated below. This disregards any additional factors introduced by the perpendicularity control on the center planes. The Gap between the bottom edge and the holes closest to it follows on the lower portion of page 270.

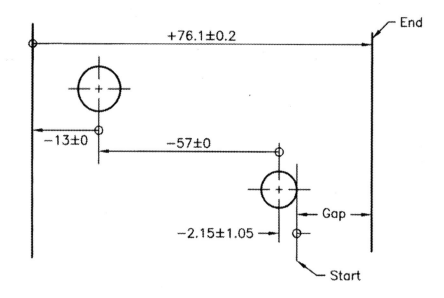

Ø4.3 (LMC) + 0.3 (Geo. Tol. at LMC) = Ø4.6 = O.B. Hole
+1.4 = Pos. Tol. of B at LMC
+0.4 = Pattern Shift (B referenced at MMC)
Ø6.4 = Hole's outer boundary plus all tolerances that apply

Ø4.0 (MMC) - 0 (Geo. Tol. at MMC) = Ø4.0 = I.B. Hole
-1.4 = Pos. Tol. of B at LMC
- 0.4 = Pattern Shift (B referenced at MMC)
Ø2.2 = Hole's outer boundary minus all tolerances that apply

Ø6.4   Ø6.4
- Ø2.2   + Ø2.2
Ø4.2 = Diff.   Ø8.6 = Sum   Divide both by 2. $\frac{Ø4.2}{2} = Ø2.1$ & $\frac{Ø8.6}{2} = Ø4.3$

So, Ø4.3±2.1 converted to a radius is R2.15±1.05
Overall dimension of part is 76.3 = O.B. and 75.9 = I.B. [76.1 - 0.2 (G.T. at LMC)]
76.3 - 75.9 = 0.4 difference and 76.3 + 75.9 = 152.2 sum
Divide both by 2. $\frac{0.4}{2} = 0.2$ and $\frac{152.2}{2} = 76.1$. So; 76.1±0.2

ANSWER - Exercise 8-2, Problem #3 (continued)

-	+	±Tol
2.15		1.05
57.00		0.00
13.00		0.00
	76.10	0.20
72.15	76.10	1.25	Totals

```
  76.10            3.95              3.95
- 72.15          - 1.25            + 1.25
  -----            ----              ----
   3.95            2.70 = MIN WALL   5.20 = MAX WALL
```

The bottom edge can be calculated using the following set of numbers:

First we must calculate the outer and inner boundary of the overall dimension in the top to bottom direction. 54.1 + 0 = 54.1 outer boundary and 53.9 - 0.2 = 53.7 inner boundary. 54.1 - 53.7 = 0.4 difference. 54.1 + 53.7 = 107.8 sum. Divide both by 2. $\frac{0.4}{2} = 0.2$ and $\frac{107.8}{2} = 53.9$. So; 53.9±0.2

-	+	±Tol
2.15		1.05
35.00		0.00
13.00		0.00
	53.9	0.20
50.15	53.9	1.25	Totals

```
  53.90            3.75              3.75
- 50.15          - 1.25            + 1.25
  -----            ----              ----
   3.75            2.5 = MIN WALL    5.00 = MAX WALL
```

ANSWER - Exercise 8-3, Problem #1

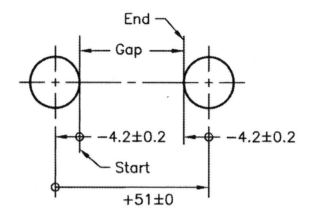

Inner Boundary of Hole = 8.0 (MMC) - 0 (Geo. Tol. at MMC) = Ø8.0
Outer Boundary of Hole = 8.4 (LMC) +0.4 (Geo. Tol. at LMC) = Ø8.8

```
  8.8 = O.B. of Hole           8.8 = O.B. of Hole
- 8.0 = I.B. of Hole         + 8.0 = I.B. of Hole
  ───────────────              ───────────────
  0.8 = Difference             16.8 = Sum
```

$\dfrac{0.8}{2} = 0.4 = \pm\text{Tolerance}$ $\dfrac{16.8}{2} = 8.4 = \text{Mean Dimension}$

Divide both by 2 to convert to radii = 4.2±0.2

−	+	±Tol
4.2		0.2
	51	0.0
4.2		0.2
8.4	51	0.4	Totals

```
  +51.0           42.6              42.6
+  -8.4        -  0.4            +  0.4
  ─────          ─────              ─────
  +42.6          42.2 = MIN GAP     43.0 = MAX GAP
```

271

ANSWER - Exercise 8-3, Problem #2

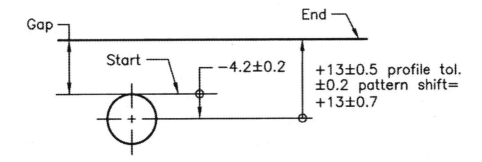

Inner Boundary of Hole = 8.0 (MMC) - 0 (Geo. Tol. at MMC) = Ø8.0
Outer Boundary of Hole = 8.4 (LMC) + 0.4 (Geo. Tol. at LMC) = Ø8.8

8.8 = O.B. of Hole - 8.0 = I.B. of Hole 0.8 = Difference	8.8 = O.B. of Hole + 8.0 = I.B. of Hole 16.8 = Sum
$\dfrac{0.8}{2} = 0.4 = \pm\text{Tolerance}$	$\dfrac{16.8}{2} = 8.4 = \text{Mean Dimension}$

Divide both by 2 to convert to radii = 4.2±0.2

-	+	±Tol	
4.2		0.2	
	13	0.7	
4.2	13	0.9	Totals

```
  13.0        8.8          8.8
-  4.2      - 0.9        + 0.9
  ----       ----         ----
   8.8        7.9 = MIN WALL    9.7 = MAX WALL
```

ANSWER - Exercise 8-3, Problem #3

Step 1:

4.3 = LMC Hole	4.0 = MMC Hole
+ 0.3 = Geo. Tol. at LMC	- 0.0 = Geo Tol. at MMC
Ø4.6 = Outer Boundary	Ø4.0 = Inner Boundary
+ 0.4 = Pattern Shift (B M)	- 0.4 = Pattern Shift
Ø5.0 = Outer Boundary with Pattern Shift	Ø3.6 = Inner Boundary with Shift
5.0 = Outer Boundary	5.0 = Outer Boundary
+ 3.6 = Inner Boundary	- 3.6 = Inner Boundary
8.6 = Sum	1.4 = Difference
$\frac{8.6}{2}$ = Ø4.3 Mean	1.4 = 0.7±Tol.
$\frac{Ø4.3}{2}$ = R2.15	$\frac{0.7}{2}$ = 0.35

So,
R2.15±0.35.

Step 2: Right Edge of Part from Datum B

76.2 = Basic Overall
- 13.0 = Basic Dimension Left Edge to B
63.2 = Basic Dimension right Edge to B

63.2±0.5 (Profile Tolerance)±0.2 (Pattern Shift-Separate Requirement)

63.2	63.2
+ 0.5	- 0.5
+ 0.2	- 0.2
63.9	62.5
63.9	63.9
+ 62.5	- 62.5
126.4	1.4
$\frac{126.4}{2}$ = 63.2	$\frac{1.4}{2}$ = 0.7

So,
63.2±0.7

ANSWER - Exercise 8-3, Problem #3 (continued)

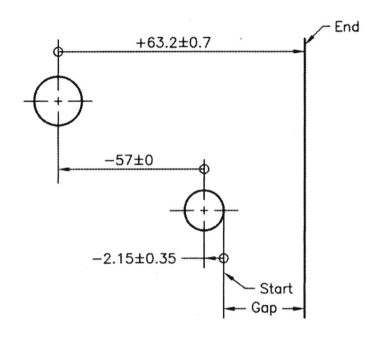

−	+	±Tol	
2.15		0.35	
57.00		0.00	
	63.20	0.70	
59.15	63.20	1.05	Totals

```
  63.20          4.05              4.05
- 59.15        + 1.05            - 1.05
  -----          ----              ----
   4.05        5.10 = MAX GAP    3.00 = MIN GAP
```

ANSWER - Exercise 8-3, Problem #3 (continued)

MIN and MAX GAP Bottom Edge

Step 1:

    Hole R2.15±0.35.

Step 2: Bottom Edge of Part from Datum B

      54 = Basic Overall
    - 13 = Basic Dimension from Top Edge to B
      41 = Basic Dimension from B to Bottom Edge

    41±0.5 (Profile Tolerance) ±0.2 (Pattern Shift-Separate Requirement)

41.0	41.0
+ 0.5	- 0.5
+ 0.2	- 0.2
41.7	40.3
41.7	41.7
+ 40.3	- 40.3
82.0	1.4
82/2 = 41	1.4/2 = 0.7

So,
    41±0.7.

ANSWER - Exercise 8-3, Problem #3 (continued)

Step 3:

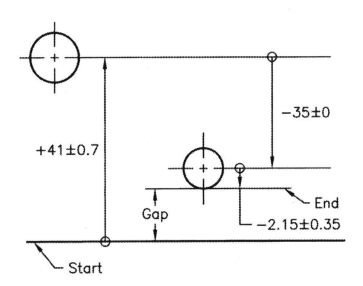

-	+	±Tol	
	41.00	0.70	
35.00		0.00	
2.15		0.35	
37.15	41.00	1.05	Totals

```
  41.00         3.85              3.85
- 37.15       + 1.05             - 1.05
  -----       ------             ------
   3.85       4.90 = MAX WALL    2.80 = MIN WALL
```

ANSWER - Exercise 8-4, Problem #4

Step 1: Outside Diameter

    2.990 = MMC   2.960 = LMC
  + .004 = Geo Tol. at MMC   − .034 = Geo Tol. at LMC
  Ø2.994 = Outer Boundary Shaft   2.926 = Inner Boundary Shaft

    2.994 = Outer Boundary   2.994 = Outer Boundary
  + 2.926 = Inner Boundary   − 2.926 = Inner Boundary
    5.920 = Sum   .068 = Difference

$$\frac{5.920}{2} = 2.960 \text{ Mean} \qquad\qquad \frac{.068}{2} = .034 \pm \text{Tol.}$$

Divide by 2 to convert to radii:

$$\frac{2.960}{2} = R1.480 \qquad\qquad \frac{.034}{2} = \pm.017$$

So,
  R1.480±.017.

Step 2: Holes

    .570 = MMC   .590 = LMC
  − .020 = Geo Tol. at MMC   + .040 = Geo Tol. at LMC
  Ø.550 = Inner Boundary   Ø.630 = Outer Boundary

    .630 = Outer Boundary   .630 = Outer Boundary
  + .550 = Inner Boundary   − .550 = Inner Boundary
    1.180 = Sum   .080 = Difference

$$\frac{1.180}{2} = .590 \text{ Mean} \qquad\qquad \frac{.080}{2} = .040 \pm \text{Tol.}$$

Divide by 2 to convert to radii:

$$\frac{\emptyset.590}{2} = R.295 \qquad\qquad \frac{.040}{2} = \pm.020$$

So,
  R.295±.020.

ANSWER - Exercise 8-4, Problem #1 (continued)

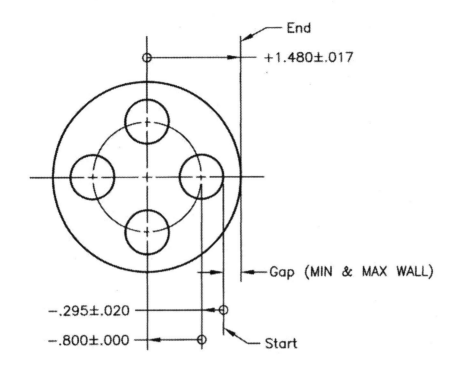

−	+	±Tol	
.295		.020	
.800		.000	
	1.480	.017	
−1.095	+1.480	.037	Totals

```
  1.480        .385             .385
- 1.095      - .037           + .037
  -----      ------           ------
   .385       .348 = MIN WALL  .422 = MAX WALL
```

278

ANSWER – Exercise 8-5, Problem #1

Step 1:

    .260 = LMC Hole
   + .050 = Geo. Tol. at LMC
   Ø.310 = Outer Boundary Hole
   + .130 =   Tol. of D at LMC
    .440
   + .130 =   Pattern Shift (D referenced at Maximum Material Boundary)
   Ø.570 = Outer Boundary Hole with all Factors

    .240 = MMC Hole
   - .030 = Geo. Tol. at MMC
   Ø.210 = Inner Boundary Hole
   - .130 =   Tol. of D at LMC
    .080
   - .130
   Ø -.050 = Inner Boundary Hole with all Factors

    +.570
   + -.050 $\frac{.520}{2}$ = .260 = Mean Dimension
    .520
and
    +.570
   - -.050 $\frac{.620}{2}$ = .310 = ± Tol.
    .620

So,
   Ø.260±.310
   R.130±.155

ANSWER – Exercise 8-5, Problem #1 (continued)

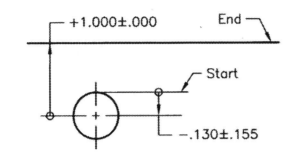

1.000	.870	.870
- .130	+ .155	- .155
.870	1.025 = MAX WALL	.715 MIN WALL
		to Datum Plane C

1/2 degree = .008 (over this part's maximum thickness)

6.100±.020 = 6.120 - 6.080 = .040 flatness tolerance for datum feature C.

.715 - .048 = .667 MIN WALL to lowest point on the surface of datum feature C.

ANSWER – Exercise 8-6, Problem #1

$$.260 = \text{LMC Hole}$$
$$+ .020 = \text{Geo. Tol. at LMC}$$
$$\emptyset.280 = \text{Outer Boundary of Hole}$$

$$\frac{\emptyset.280}{2} = \text{Radius of Hole} = .140$$

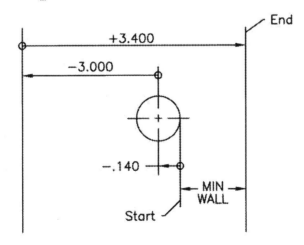

3.400	.260
- 3.140	- .009
.260	.251 = MIN WALL

ANSWER – Exercise 8-7, Problem #1

Step 1:

　　.375 = LMC Hole
　+.035 = Geo. Tol. at LMC
　Ø.410 = Outer Boundary
　+ .400 = Pattern Shift (Dμ)
　Ø.810 = Outer Boundary Hole with Pattern Shift
　+ .400 = Position Tol. of D
　Ø1.210 = Outer Boundary Hole with all Facets

　R.605

Step 2:

　4.990 = LMC Outside Diameter
　- .040 = Geo. Tol. at LMC
　Ø4.950 = Inner Boundary of Outside Diameter

　$\dfrac{Ø4.950}{2} = R2.475$

Step 3:

　1.600
　x $\sqrt{2}$　 = 1.4142
　Ø2.263
　R1.1315

ANSWER – Exercise 8-7, Problem #1 (continued)

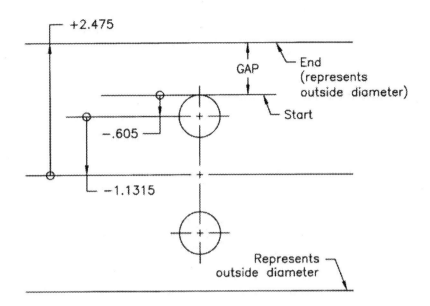

−	+
.6050	
1.1315	
	2.475
1.7365	2.475	Totals

```
  2.4750
- 1.7365
  ─────
   .7385 = MIN WALL
```

ALTERNATE METHOD ANSWER – Exercise 8-7, Problem #1 (continued)
[Double check using diameter method.]

Step 1: Shaft

```
  4.990 = Shaft
-  .040 = Geo. Tol. at LMC
  4.950 = Inner Boundary Shaft
```

Step 2: Bolt Square converted to Bolt Circle

$$\frac{1.600 \times \sqrt{2}}{\varnothing 2.263}$$

Step 3:

```
  4.950 = Inner Boundary Shaft
- 2.263 = Bolt Circle Diameter
  2.687 = Material with Hole
```

Step 4: Hole

```
  .375 = LMC Hole
+ .035 = Geo. Tol. at LMC
 Ø.410 = Outer Boundary Hole
```

Step 5:

```
  2.687
-  .410
  2.277
```

$$\frac{2.277}{2} = 1.1385$$

Step 6:

```
  .400 = Accumulated Error from D (Position Tol.)
+ .400 = Pattern Shift of 4 Holes Due to D µ
 Ø.800
```

$$\frac{.800}{2} = R.400$$

Step 7:

```
  1.1385
-  .4000
   .7385 = MIN WALL
```

. . . matches answer using Tolerancing Stack-Up approach.

ANSWER – Exercise 8-8, Problem #1

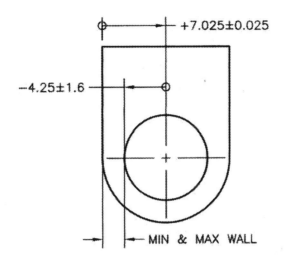

```
  8.5 = LMC Hole                    8.3 = MMC
+ 3.2 = Geo. Tol. at LMC          - 3.0 = Geo. Tol. at MMC
 11.7 = Outer Boundary of Hole     5.3 = Inner Boundary of Hole


  11.7 = Outer Boundary of Hole    11.7 = Outer Boundary of Hole
+  5.3 = Inner Boundary of Hole   - 5.3 = Inner Boundary of Hole
  17.0 = Sum                       6.4 = Difference
```

$\dfrac{17}{2} = 8.5$ Mean Dimension $\dfrac{6.4}{2} = 3.2 \pm$ Tol.

Ø8.5±3.2 divided by 2 to convert to a radius equals R4.25±1.6

Datum feature B;   14.1 = Outer Boundary of B   14.1 = Outer Boundary of B
                 + 14.0 = Inner Boundary of B   - 14.0 = Inner Boundary of B
                   28.1 = Sum   0.1 = Difference

$\dfrac{28.1}{2} = 14.05$ and $\dfrac{0.1}{2} = 0.05$ equals 14.05±0.05

14.05±0.05 divided by 2 to convert to a radius equals R7.025±0.025

ANSWER – Exercise 8-8, Problem #1 (continued)

	−	+	±Tol
		7.025	0.025
	4.250		1.600
	−4.250	+7.025	1.625	Totals

```
   7.025          2.775              2.775
 − 4.250        − 1.625            + 1.625
   -----          -----              -----
   2.775        1.150 = MIN WALL   4.400 = MAX WALL
```

This would seem to be the end of the analysis, but in this case another look at the part is necessary. With a feature of size as the primary datum feature, an odd situation is possible. In our analysis above, we divided every factor into radii and calculated the MIN and MAX walls from the radial factors. This is sufficient for the MAX wall, but not for the MIN wall. All of the size tolerance can gravitate to one edge of datum feature B. The illustration below shows the result.

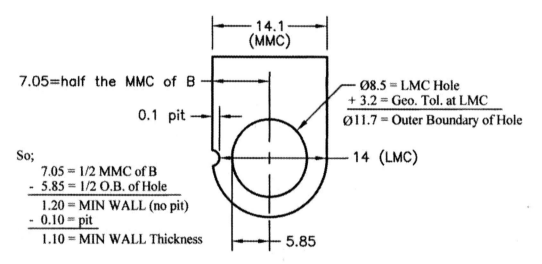

So;
```
  7.05 = 1/2 MMC of B
− 5.85 = 1/2 O.B. of Hole
  ----
  1.20 = MIN WALL (no pit)
− 0.10 = pit
  ----
  1.10 = MIN WALL Thickness
```

This second analysis just shows the importance of judgment and imagination in any analysis and why tolerance stack-up analysis is so hard to get right when pat routines are followed without every possibility being considered.

ANSWER - Exercise 9-1

Part #4

Step 1: Outside Diameter

      263.0 = MMC
     + 0.2 = Geo. Tol.
     Ø263.2 = Outer Boundary

$$\frac{263.2}{2} = R131.6 \text{ Shaft}$$

Step 2: Clearance Hole

     Ø9.0 = LMC

$$\frac{9.0}{2} = R4.5 \text{ Hole}$$

Part #3

Step 1: Mounted Screw (in threaded hole)

      7.76 = LMC Screw
     - 0.40 = Geo. Tol.
      7.36 = Inner Boundary Screw
     - 0.06 = Pattern Shift due to Dµ Reference
     Ø7.30 = Inner Boundary with Pattern Shift

$$\frac{7.3}{2} = R3.65 \text{ Shaft}$$

Step 2:

     Ø79.9 = LMC of D

$$\frac{Ø79.9}{2} = R39.95 \text{ Shaft}$$

Part #2

Step 1:

     Ø80.08 = LMC of D
    + 0.20 = Position Tol. of D
     Ø80.28 = Outer Boundary of D

$$\frac{80.28}{2} = R40.14 \text{ Hole}$$

Step 2:

     Ø109.00 = LMC of E

$$\frac{109}{2} = R54.5 \text{ Shaft}$$

Part #1

Step 1: Central Hole

     111.00 = LMC Hole
   + 1.20 = Geo. Tol. at LMC
     112.20 = Outer Boundary Hole

$$\frac{112.2}{2} = R56.1 \text{ Hole}$$

MMC of C = 264.6 = R132.3

ANSWER - Exercise 9-1 (continued)

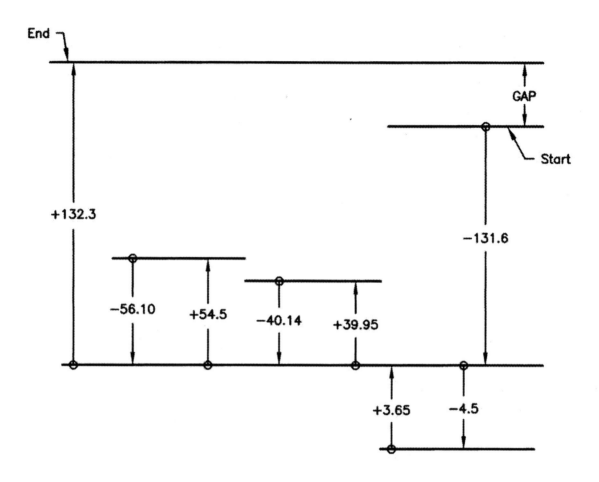

$$+230.40$$
$$+ \ -232.34$$
$$\overline{-1.94}$$ 1.94 = MAX. Interference

ANSWER - Exercise 10-1

This problem is solved by doing the following proportional calculation:

$$\frac{.020}{36} = \frac{x}{16}$$

$$.020 \cdot 16 = 36x$$

$$.32 = 36x$$

$$\frac{.32}{36} = \frac{36x}{36}$$

$$\frac{.32}{36} = x$$

$$.0089 = x$$

So, the increase is .0089 and the virtual condition for the hole in Part #3 is Ø.2400 + .0089 = Ø.2489. With the geometric tolerance stated as .000 at MMC, the MMC will also be Ø.2489.

ANSWER - Exercise 11-1

To answer the Exercise 11-1 question on this methodology, we used the illustration from Chapter 6 where in a fixed fastener 2-part assembly we calculated the MIN GAP for the lower left. To do this, we had to choose the correct route to follow, convert the dimensions to equal bilateral with plus or minus tolerances, then calculate the gap. We used basic dimensions that had tolerances expressed as zero and slots and tabs whose tolerances included both size and position. Still, in the end, we came up with a series of plus and minus tolerances that we used to calculate the minimum gap. Now we will use that illustration to show how we can convert to statistical tolerances with the same methods we used on the previous examples.

−	+	±	± squared	Type
24.00		0.200	0.0400	Wall
	190.60	0.000	0.0000	Basic Dim
12.19		0.110	0.0121	Slot
	11.97	0.110	0.0121	Tab
	114.20	0.000	0.0000	Basic Dim
273.00		1.400	1.9600	Overall Dim
309.19	316.77	1.820	2.0242	Totals

$$\begin{array}{r} 316.77 \\ - 309.19 \\ \hline +7.58 \end{array}$$ = difference between (technically the sum of) the positive and negative mean dimensions

2.0242 = Sum of the Squares

$\sqrt{2.0242} = 1.4227438$ = Square Root of the Sum of the Squares (RSS)

$\dfrac{1.4227438 \text{ (tol. likely to be consumed)}}{1.82 \text{ (arithmetically calculated tol.)}} = 0.7817273$

So, 1.4227438 is about 78% of 1.82.

$\dfrac{1}{0.7817273} = 1.2792184$ and $1.2792184 \times 1.4227438 = 1.82$ (the originally calculated arithmetic tolerance)

ANSWER - Exercise 11-1 (continued)

So, the arithmetically calculated 100% tolerance allows a minimum gap that is 7.58 - 1.82 = 5.76.

The statistically calculated assembly tolerance predicts a minimum gap that is 7.58 - 1.42 (rounded off to two decimal places). This 1.42 tolerance is the amount of tolerance likely to be consumed in a natural bell curved distribution of manufactured parts for this assembly. So, if we want to consume ±1.82 tolerance, the piece part tolerance should be increased by 128% (rounded off from 127.92184%).

So, the ±0.11 tolerance for the Slot and the Tab becomes 1.28 x 0.11 = 0.1408 or a tolerance for each that is ±0.141 (rounded off to three decimal places). The tolerance for the Wall becomes 1.28 x 0.200 = ±0.256. The tolerance for the Overall dimension becomes: 1.28 x 1.4 = ±1.792. This is the answer to the problem.

Since statistically the tolerance originally assigned would not be fully consumed, the minimum gap calculation given originally as 7.58 - 1.82 = 5.76 MIN GAP becomes a statistical probability within ±3 sigma of consuming only 7.58 - 1.42 = 6.16 MIN GAP.

Given the newly assigned statistically calculated tolerances, with each piece part given a statistically calculated tolerance, we have a mathematical possibility of a MIN GAP that is 7.58 minus the sum of the statistical tolerances. They are: 0.141 (Slot) + 0.141 (Tab) + 0.256 (Wall) + 1.792 (Overall dim.) = ±2.33.

$$\begin{array}{r}0.141 \text{ (slot)}\\+\ 0.141 \text{ (Tab)}\\+\ 0.256 \text{ (Wall)}\\\underline{+\ 1.792 \text{ (Overall dim.)}}\\\pm 2.33\end{array}$$

And it leaves a MIN GAP of 7.58 - 2.33 = 5.25. So, arithmetically we might have a MIN GAP that is 5.25, but this is highly unlikely.
- The Slot and Tab statistical tolerances of 0.141 when squared become 0.019881 each.
- The Wall statistical tolerance of 0.256 when squared becomes 0.065536.
- The Overall dimension statistical tolerance of 1.792 when squared becomes 3.211264.

- When added, these tolerances are:

$$\begin{array}{r}0.019881\\+\ 0.019881\\+\ 0.065536\\\underline{+\ 3.211264}\\3.316562\end{array}$$

The square root of 3.316562=1.82=Square Root of the Sum of the Squares –RSS (rounded off).

ANSWER - Exercise 11-1 (continued)

So, we have again shown by the RSS (Root Sum Square) formula that even though the statistical assembly tolerance (±2.33) is greater (by 128%) than originally calculated arithmetically as ±1.82, the amount of tolerance likely to be consumed is still only ±1.82.

Now we can reintegrate the Statistical Tolerance back into the assembly for the slot and the tab. The ±0.11 tolerance for the slot and tab shown in the problem becomes 1.28 x 0.11 = 0.1408 or a tolerance for each that is ±0.1408.

To reintegrate these tolerance into a combination of size and geometric tolerance, the following steps are performed:

±0.1408 for the slot:

1) 12.19 x 2 = 24.38

2) 0.1408
   × 2
   0.2816 = Tol

3) 24.3800   24.3800
   - 0.2816   + 0.2816
   24.0984 = I.B.   24.6616 = O.B.

4) 0.1 = Original Geo. Tol. at MMC from page 6-30 so;   1.28 = Increase
                                                        × 0.10 = Geo. Tol. at MMC
                                                        0.128 = new ⊕ Tol. at MMC ⟨ST⟩

   | ⊕ | 0.128 Ⓜ ⟨ST⟩ | A | B | C |

5) 0.22 = Geo. Tol. at LMC so;   1.28 = Increase
                                 × 0.22 = Geo. Tol. at LMC
                                 0.2816 = new Geo. Tol. ⊕ Tol. at LMC ⟨ST⟩

6) Add 0.128 to I.B. 24.0984
   24.0984
   + 0.1280
   24.2264 = MMC of the slot

7) Subtract 0.2816 from O.B. 24.6616
   24.6616
   - 0.2816
   24.3800 = LMC of the slot

8) ⌀24.2264 - 24.3800 ⟨ST⟩

   | ⊕ | 0.128 Ⓜ ⟨ST⟩ | A | B | C |

ANSWER - Exercise 11-1 (continued)

To reintegrate these tolerances into a combination of size and geometric tolerance, the following steps are performed:

±0.1408 for the tab:

1) 11.97 x 2 = 23.94

2) 0.1408
   $\times$ 2
   0.2816 = Tol

3) 23.9400   23.9400
   - 0.2816   + 0.2816
   23.6584 = I.B.   24.2216 = O.B.

4) 0.1 = Original Geo. Tol. at MMC from page 6-30 so;   1.28 = Increase
   $\times$ 0.10 = Geo. Tol. at MMC
   0.128 = new ⊕ Tol. at MMC ⟨ST⟩

   | ⊕ | 0.128 Ⓜ ⟨ST⟩ | D | E | F |

5) 0.22 = Geo. Tol. at LMC so;   1.28 = Increase
   $\times$ 0.22 = Geo. Tol. at LMC
   0.2816 = new Geo. Tol. ⊕ Tol. at LMC ⟨ST⟩

6) Subtract 0.128 from O.B. 24.2216
   24.2216
   - 0.1280
   24.0936 = MMC of the tab

7) Add 0.2816 to the I.B. 23.6584
   23.6584
   + 0.2816
   23.9400 = LMC of the tab

8) 23.9400-24.0936 ⟨ST⟩

   | ⊕ | 0.128 Ⓜ ⟨ST⟩ | D | E | F |

ANSWER – Exercise 11-1 (continued)

Using the Simpler Methodology to Keep the Same Size Mean Dimension Instead of the Same Functional Boundary Mean

SLOT-With this method, the original size dimension on the slot of 24.32 is kept. The original size tolerance of ±0.06 is multiplied by 1.28 to get 0.0768. So, the new size of the slot is **24.32±0.0768 Statistical Tolerance**. The original position tolerance of 0.1 at MMC is multiplied by 1.28 to become **0.128 at MMC Statistical Tolerance**.

TAB-With this method, the original size dimension on the tab of 24 is kept. The original size tolerance of ±0.06 is multiplied by 1.28 to become 0.0768. So, the new size of the slot is **24±0.0768 Statistical Tolerance**. The original position tolerance of 0.1 at MMC is multiplied by 1.28 to become **0.128 at MMC Statistical Tolerance**.

ANSWER – Exercise 11-2

–	+	±	± squared	Type
3		0.7	0.49	Hole #1
	3	0.5	0.25	Pin
3		0.5	0.25	Hole #3
3		0.5	0.25	Hole #4
	3	0.5	0.25	Pin
3		0.7	0.49	Hole #2
12	6	3.4	1.98	Totals

$1.5 \sqrt{1.98} = 2.110687$

$\dfrac{2.110687 \text{ (tol. likely to be consumed)}}{3.4 \text{ (arithmetically calculated tol.)}} = 0.6207902$

So, 2.110687 is about 62% of 3.4.

$\dfrac{1}{0.6207902} = 1.6108501$ and $1.6108501 \times 2.110687 = 3.4$ (the originally calculated arithmetic tolerance)

ANSWER – Exercise 11-2 (continued)

So, we will increase the tolerances for the pertinent features by 161%.

So, the ±0.7 tolerance for Hole #1 and Hole #2 becomes 1.61 x 0.7 = 1.127 or a tolerance for each that is ±1.127. The tolerance for Hole #3, Hole #4 and both Pins becomes 1.61 x 0.5 = ±0.805. <u>This is the answer to the problem.</u>

To reintegrate these tolerances into a combination of size and geometric tolerance, the following steps are performed:

±1.127 for Holes #1 and #2:

1) 3 x 2 = Ø6

2) 1.127
   x 2
   ─────
   2.254 = Tol

3) 6.000   6.000
   - 2.254   + 2.254
   ─────   ─────
   3.746 = I.B. 8.254 = O.B.

4) 0.4 = Original Geo. Tol. at MMC so;   1.61 = Increase
                                         x 0.4 = Geo. Tol. at MMC
                                         ─────
                                         0.644 = new ⊕ Tol. at MMC ⟨ST⟩

   | ⊕ | Ø 0.644 Ⓜ ⟨ST⟩ | A | B | C |

5) 1.4 = Geo. Tol. at LMC so;   1.61 = Increase
                                x 1.4 = Geo. Tol. at LMC
                                ─────
                                2.254 = new Geo. ⊕ Tol. at LMC ⟨ST⟩

6) Add 0.644 to I.B. 3.746
   3.746
   + 0.644
   ─────
   4.390 = MMC

7) Subtract 2.254 from O.B. 8.254
   8.254
   - 2.254
   ─────
   6.000 = LMC

8) Ø4.39 - 6.00 ⟨ST⟩

   | ⊕ | Ø 0.644 Ⓜ ⟨ST⟩ | A | B | C |

ANSWER – Exercise 11-2 (continued)

±0.805 for Holes #3 and #4;

1) 3 x 2 = Ø6

2) $\quad$ 0.805
   $\quad$ x 2
   $\quad$ ─────
   $\quad$ 1.610 = Tol

3) $\quad$ 6.000 $\qquad\qquad$ 6.000
   $\quad$ - 1.610 $\qquad\quad$ + 1.610
   $\quad$ ─────── $\qquad$ ───────
   $\quad$ 4.390 = I.B. $\qquad$ 7.610 = O.B.

4) 0 = Original Geo. Tol. so; 0 x 1.61 = 0

   | ⊕ | Ø 0 Ⓜ ⟨ST⟩ | A | B | C |

5) 1mm = Geo. Tol. at LMC so; $\qquad$ 1.61
   $\qquad\qquad\qquad\qquad\qquad\qquad\qquad$ x 1
   $\qquad\qquad\qquad\qquad\qquad\qquad\qquad$ ─────
   $\qquad\qquad\qquad\qquad\qquad\qquad\qquad$ 1.61 = new Geo. Tol. ⊕ Tol. at LMC ⟨ST⟩

6) Add 0 to I.B. 4.390 so; new MMC = 4.390

7) Subtract 1.61 from O.B. 7.610
   $\quad$ 7.610
   $\quad$ - 1.610
   $\quad$ ───────
   $\quad$ 6.000 = LMC

8) Ø4.39 - 6.00 ⟨ST⟩

   | ⊕ | Ø 0 Ⓜ ⟨ST⟩ | A | B | C |

Had we used the RSS formula <u>without the correction factor</u> of 1.5, we would have found the following;

$$\sqrt{1.98} = 1.4071247$$

$$\frac{1.4071247 \text{ (tol. likely to be consumed)}}{3.4 \text{ (arithmetically calculated tol.)}} = 0.4138602$$

so, 1.4071247 is about 41% of 3.4.

$$\frac{1}{0.4138602} = 2.4162748 \quad \text{and} \quad \begin{array}{r} 2.4162748 \\ \times\ 1.4071247 \\ \hline 3.4 \text{ (the original calculated arithmetic tolerance)} \end{array}$$

So, had we used the RSS formula without the 1.5 correction factor, the tolerance for the pertinent features would have been increased by 242% instead of 161%.

ANSWER – Exercise 11-2 (continued)

Using the Simpler Methodology to Keep the Same Size Mean Dimension Instead of the Same Functional Boundary Mean

Holes 1 & 2-With this method, the original size mean dimension on the holes of Ø5.5 is kept. The original size tolerance of ±0.5 is multiplied by 1.61 to get 0.805. So, the new size of the slot is **Ø5.5±0.805 Statistical Tolerance** using the Root Sum Square method with a safety factor of 1.5. The original position tolerance of Ø0.4 at MMC is multiplied by 1.61 to become **Ø0.644 at MMC Statistical Tolerance** using the Root Sum Square method with a safety factor of 1.5.

Holes 3 & 4-With this method, the original size mean dimension on the holes of Ø5.5 is kept. The original size tolerance of ±0.5 is multiplied by 1.61 to get 0.805. So, the new size of the slot is **Ø5.5±0.805 Statistical Tolerance** using the Root Sum Square method with a safety factor of 1.5. Since the original position tolerance for holes 3 & 4 was zero at MMC, mulitplying it by anything leaves it unchanged. So, the position tolerance is still **Ø0 at MMC**.

APPENDIX A

On-Site Training Programs and Training Materials Available

- ## Courses for On-Site Training at your facility

- ## Training Materials - written by James D. Meadows
  ### Textbooks & Workbooks
  - *Geometric Dimensioning and Tolerancing-Applications, Analysis & Measurement [per the ASME Y14.5-2009 and 1994 Standards]* NEW! (...and the differences between them)
  - Workbook and Answerbook for *Geometric Dimensioning and Tolerancing [per the ASME Y14.5-2009 and 1994 Standards]* NEW!
  - *'NEW RULES' in GD&T [per ASME Y14.5-2009]* NEW!
  - *Geometric Dimensioning and Tolerancing in 2007 (per ASME Y14.5M-1994)*
  - *Geometric Dimensioning and Tolerancing in 2007 Workbook and Answerbook (per ASME Y14.5M-1994)*
  - *Differences and Similarities between ASME and ISO Dimensioning and Tolerancing Standards*
  - *Geometric Dimensioning and Tolerancing: Applications and Techniques for Use in Design, Manufacturing and Inspection (per ASME Y14.5M-1994]*
  - *Geometric Dimensioning and Tolerancing: Workbook and Answerbook (per ASME Y14.5M-1994)*
  - *Measurement of Geometric Tolerances in Manufacturing*
  - *Tolerance Stack-Up Analysis 2nd Edition [per ASME Y14.5]*

  ### GD&T Training Series [DVD]
  - *'NEW RULES' in GD&T DVD Training Series [per ASME Y14.5-2009]* NEW!
  - *Applications-Based Geometric Dimensioning and Tolerancing DVD Training Series [per ASME Y14.5M-1994]*

- ## Engineering Newsletter [E-Newsletter at www.geotolmeadows.com]

ON-SITE TRAINING

We will come on-site to provide a wide variety of Geometric Dimensioning and Tolerancing training and support programs, both "packaged" and customized and, if requested, will incorporate your processes and products. Our GDT training personnel have spent years working in all facets of design, manufacturing and quality, so your staff can have complete confidence in the guidance that they receive.

We offer course lengths from 2 1/2 to 5 days in length to meet your training needs and time constraints. We provide "applications-based" courses on the principles contained in the NEW ASME Y14.5-2009 or the ASME Y14.5M-1994 standard on Geometric Dimensioning and Tolerancing. This training can be tailored to include your product lines, designs, and manufacturing processes.

-All course levels are *'applications-based'* training-

LEVEL 1

●Principles and Applications of Geometric Dimensioning and Tolerancing

ASME Y14.5 training for all job categories; **the foundation of GD&T principles for all (advanced) level 2 courses**. An introductory, but comprehensive, applications-based training program for design, manufacturing, quality, inspection, purchasing staff and managers. The goal of this course is not only to give the participants a thorough knowledge of GD&T techniques, but through the use of discussion, historical data and application problems, the ability to apply these techniques to their product lines with a great deal of confidence and a perspective on how each symbol and concept compares for strengths and weaknesses, capabilities, advantages and disadvantages.

LEVEL 2 COURSES *[provide advanced 'applications' training in various aspects of GD&T]*

●Advanced Geometric Dimensioning and Tolerancing Applications

This course is directed to design, drafting and engineering personnel or anyone wishing to gain an advanced knowledge of applying geometric symbology, principles and rules to product designs. This advanced training can be **tailored to fit your product lines**. It is the most comprehensive GD&T design applications course offered anywhere.

This course applies to products made of all material types and that are produced using any manufacturing procedure. Rigid products that mate, rotate, are cylindrical, rectangular, curved or oddly configured are covered in great depth and detail. Also covered are parts that are prone to free-state variation, that can "flex" and deform when you try to measure or assemble them. Learn how to prevent problems and minimize disagreement through the proper use of GD&T.

This course is targeted at those who have already been trained in the basics of Geometric Dimensioning and Tolerancing. Even the most advanced practitioners will learn more than they ever thought possible. Some of the general topic categories covered are: GD&T Refresher, Advanced GD&T Concepts, Tolerance Stack-Up Analysis, Statistical Tolerancing, CMM Measurement, Variables Data Collection and Analysis, Functional Gage and Fixture Design, Dimensioning and Tolerancing. Based on the most comprehensive GD&T textbook ever written by a single author, *Geometric Dimensioning and Tolerancing (per ASME Y14.5-2009)*, this course has the unprecedented ability to cover almost every facet of tolerancing per the ASME Y14.5-2009 and 1994 Standards. Unlike any before it, this Advanced GD&T course covers almost every tolerancing-related topic--time is the only limiting factor. The course, and textbook it is based on, have it all.

●Tolerance Stack-Up Analysis

Learn methods to analyze tolerance stack-ups using plus and minus and geometric tolerancing. This course covers the analysis of small assemblies, large assemblies and single parts. It includes both worst case analysis and statistical analysis. All pertinent factors are discussed and practiced. Procedures show how to include or exclude pertinent factors, how to perform several methods of a circuit analysis and how to calculate trigonometric, proportional and statistical considerations. It teaches the logic and the methodology developed over decades and used by the military, government projects (DOD, DOE, NASA, etc.) and all facets of private industry.

- **Dimensioning and Tolerancing of Functional Gages and Fixtures and Variables Data Collection and Analysis**

Learn to design, dimension and tolerance gages and fixtures to inspect geometric tolerances per the ASME Y14.43 standard--of which James D. Meadows is chairman. Attendees will also gain a thorough understanding of the interpretation and inspection ramifications of drawings dimensioned and toleranced per ASME Y14.5M, will learn to apply variables data collection methods and datum establishment by CMMs and other inspection tools, and will learn techniques to analyze collected variables data.

...OR A COMBINATION OF THE ABOVE LISTED COURSES

Course outlines presented herein are generic, but they **can be tailored specifically** to your needs--your product line and design, manufacturing and inspection requirements using your engineering drawings as a basis for your on-site training to emphasize the issues you are facing.

COURSE PREREQUISITES: The Level I course offered above (or an equivalent Basic GD&T course) are the only prerequisites for any Level 2 courses. **[For a breakdown of course outlines, see @ www.geotolmeadows.com.]**

Copyrighted books and course materials written by James D. Meadows:

GEOMETRIC DIMENSIONING AND TOLERANCING
Applications, Analysis & Measurement
(per ASME Y14.5-2009)
March 2009
574 pgs, hardcover, illus.
ISBN: 0-9714401-6-6

WORKBOOK and ANSWERBOOK for GEOMETRIC DIMENSIONING and TOLERANCING
(per ASME Y14.5-2009)
March 2009
367 pgs, softcover, illus.
ISBN: 0-9714401-7-4

'NEW RULES' in GD&T
(per ASME Y14.5-2009)
March 2009
160 pgs, softcover, illus.
ISBN: 0-9714401-8-2

'NEW RULES' in GD&T
(per ASME Y14.5-2009)
DVD Training Series
March 2009
Includes DVD set & 'NEW RULES' text
ISBN: 0-9714401-9-0

GEOMETRIC DIMENSIONING AND TOLERANCING IN 2007
(per ASME Y14.5M-1994)
February 2007
560 pgs, illus.
ISBN: 0-9714401-2-3

GEOMETRIC DIMENSIONING AND TOLERANCING IN 2007
Workbook and Answerbook
(per ASME Y14.5M-1994)
292 pgs, illus.
ISBN: 0-9714401-5-8

TOLERANCE STACK-UP ANALYSIS
[for Plus and Minus and Geometric Tolerancing]
Second Edition, Published 2010
300 pages, illus. [per the ASME Y14.5-2009 and 1994 standards]
ISBN: 978 0971 440142

DIFFERENCES AND SIMILARITIES BETWEEN ASME AND ISO DIMENSIONING AND TOLERANCING STANDARDS
December 2006 English Edition
80 pgs, illus.
ISBN: 0-9714401-3-1

GEOMETRIC DIMENSIONING AND TOLERANCING
Workbook and Answerbook
(per ASME Y14.5M-1994)
340 pgs, illus.
ISBN: 0-8247-0076-7

GEOMETRIC DIMENSIONING AND TOLERANCING
Applications and Techniques for Use in Design, Manufacturing and Inspection
(per ASME Y14.5M-1994)
624 pgs, illus.
ISBN: 0-8247-9309-9

MEASUREMENT OF GEOMETRIC TOLERANCES IN MANUFACTURING
(per ASME Y14.5M-1994)
496 pgs; illus.
ISBN 0-8247-0163-1

APPLICATIONS-BASED GD&T DVD TRAINING SERIES I
(per ASME Y14.5M-1994)
Complete Set: 12 DVD GD&T Lesson Series (14 hours)
plus Book *and* Workbook/Answerbook

APPLICATIONS-BASED GD&T DVD TRAINING SERIES II
(per ASME Y14.5M-1994)
"BASICS" Set: 4 DVD GD&T Lesson Series (4 hours)
plus Book *and* Workbook/Answerbook

 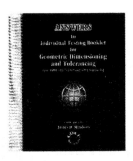 

INDIVIDUAL TESTING BOOKLET
for Geometric Dimensioning and Tolerancing
[per the ASME Y14.5-2009 and 1994 Standards]
ISBN: 978-0-615-36104-8

ANSWERS to Individual Testing Booklet
for Geometric Dimensioning and Tolerancing
[per the ASME Y14.5-2009 and 1994 Standards]
ISBN 978-0-615-36101-1

TESTING PACKAGE
10 Individual Testing Booklets
1 ANSWERS for Individual Testing Booklet
* Get 1 Testing Booklet and 1 ANSWER booklet
FREE when ordering the Testing Package

For more information, please visit www.geotolmeadows.com or call (615) 824-8644